W0230461

In memoriam
Ludwig F. Audrieth

Coordination Chemistry in Non-Aqueous Solutions

V. Gutmann

1968

Springer-Verlag

Wien · New York

Dipl.-Ing. Dr. techn. (Vienna), Ph. D. (Cantab.), Sc. D. (Cantab.) VIKTOR GUTMANN
Professor of Inorganic Chemistry, Technische Hochschule Wien, Austria

With 10 Figures

ISBN-13: 978-3-7091-8196-6   e-ISBN-13: 978-3-7091-8194-2
DOI: 10.1007/978-3-7091-8194-2

All rights reserved
No part of this book may be translated
or reproduced in any form
without written permission from Springer-Verlag

© 1968 by Springer-Verlag/Wien

Softcover reprint of the hardcover 1st edition 1968

Library of Congress Catalog Card Number 68-13490

Title No. 9226

Preface

Considerable attention has been focussed on non-aqueous chemistry in the last decade and this situation has arisen no doubt from a realization of the vast application of this branch of chemistry. Within this field much energetic work has been channelled into the determination of the coordination chemistry of transition metals in these solvent systems. Elaborate experimental techniques have been developed to discover, in particular, the magnetic and spectral properties of complex compounds, and the theoretical background of such systems has been expanded to corroborate, as far as possible, the experimental results.

This text has, however, a different bias from many books currently available on this branch of chemistry, and is designed to be a survey of known facts on many of the non-aqueous solvents currently in use mainly in the field of halogen chemistry, together with a discussion of these facts in the light of accepted principles. As such, it is hoped to close a gap in the literature of which many workers and advanced students in this field will be aware. The treatment is meant to be selective rather than completely comprehensive and must unevitably reflect some of the special interests of the author.

The solvents are classified from the point of view of coordination chemistry, as either "acceptors" or "donors" and reactions occurring in the solutions are related to the coordinating properties of the solvents. For this reason, the various physicochemical methods in use in the authors own laboratories, and the semi-quantitative data thereby available, have been utilized to establish the coordination forms of transition metals in various non-aqueous solutions and to correlate the results with the various solvent properties. The reader will, however, not find in the text discussion of ligand field theory, analytical applications of some of the reactions mentioned, or the electrochemistry (including polarography) of the solutions, because these aspects of non-aqueous chemistry, although of extense interest, are outside the scope of this volume.

It has been attempted, however, to present apart from the treatment of principles of coordination chemistry in non-aqueous solvents, a description of a large variety of solvents and of many reactions occurring in their solutions. More than 850 references have been included. It is hoped, in this way to approach both the advanced student and, in fact, every chemist interested in solution chemistry.

I wish to thank all those who have encouraged me in writing this book, in particular to the late Professor LUDWIG F. AUDRIETH, to whom this book is dedicated and to Professor HARRY H. SISLER, University of Florida, Gainesville/Fla. I wish also to thank Dr. J. FROST for his valuable assistance in preparing this text in English and Dipl.-Ing. M. BERMANN for proof reading and indexing.

Vienna, Fall 1967   V. GUTMANN

Contents

Page

Chapter I

General

1. Introduction 1
2. Classification of Solvents 1
3. Physical Properties 4
4. Acids and Bases (Acidic and Basic Function) 5
   A. Classical Definition 5
   B. The Protonic Concept 6
   C. The Solvent-System Concept 7
   D. The Ionotropic Definitions 8
   E. The Lewis Concept 9
   F. Hard and Soft Acids and Bases 10
5. Techniques with Non-Aqueous Solvents 11

Chapter II

Principles of Coordination Chemistry in Non-Aqueous Solutions

1. Donor Solvents and Acceptor Solvents 12
2. Coordinating Properties of Solvents 12
   A. Donor Strength 12
   B. The Donor Number 19
   C. Prediction of Donor-Acceptor Interactions 20
   D. Hydrogen Bonding 21
   E. Steric Factors 23
   F. Solubilities 23
3. Coordination Equilibria in Solution 24
   A. Formation of a (Non-solvated) Complex Anion 25
   B. Ionization and Dissociation of a Solvate Complex 28
   C. Autocomplex Formation 30
4. Solvation and Donor Properties 31

Chapter III

Coordination Chemistry in Proton-containing Donor Solvents

1. General Properties of Proton-containing Solvents 35
2. Liquid Ammonia 38
3. Hydrazine 48
4. Hydrogen Sulphide 49
5. Formamide and Acetamide 51
6. Formic Acid and Acetic Acid 53
7. Alcohols 56

Chapter IV

Proton-containing Acceptor Solvents

1. Introduction 59
2. Hydrogen Fluoride 60

3. Liquid Hydrogen Chloride, Hydrogen Bromide and Hydrogen Iodide 64
4. Liquid Hydrogen Cyanide 67
5. Sulphuric Acid 69
6. Nitric Acid and Phosphoric Acid 75
7. Fluorosulphuric Acid, Chlorosulphuric Acid, Difluorophosphoric Acid and Disulphuric Acid 77

Chapter V

Proton-free Acceptor Solvents

1. Covalent Oxides 80
   A. Liquid Sulphur Dioxide 80
   B. Liquid Dinitrogen Tetroxide 84
2. Covalent Fluorides 86
   A. Bromine(III) fluoride and Chlorine(III) fluoride 87
   B. Iodine(V) fluoride 91
   C. Arsenic(III) fluoride 92
3. Covalent Chlorides 94
4. Covalent Bromides 98
5. Molten Iodine 101

Chapter VI

Oxyhalide Solvents

1. Oxyhalides with Low Donor Numbers 104
2. Oxyhalides with Medium Donor Numbers 111

Chapter VII

Certain Donor Solvents

1. 1,2-Dichloroethane ($DN_{SbCl_5} = 0.1$) 127
2. Nitromethane (NM) ($DN_{SbCl_5} = 2.7$) and Nitrobenzene (NB) ($DN_{SbCl_5} = 4.4$) 127
3. Acetic Anhydride (AA) ($DN_{SbCl_5} = 10.5$) 129
4. Acetonitrile (AN) ($DN_{SbCl_5} = 14.1$) 131
5. Sulpholane (Tetramethylenesulphone) ($DN_{SbCl_5} = 14.8$) 141
6. Propanediol-1,2-carbonate (PDC) ($DN_{SbCl_5} = 15.1$) 142
7. Acetone ($DN_{SbCl_5} = 17.0$) 144
8. Ethyl Acetate ($DN_{SbCl_5} = 17.1$) 146
9. Diethylether ($DN_{SbCl_5} = 19.2$) 147
10. Trimethyl Phosphate (TMP) ($DN_{SbCl_5} = 23$) 148
11. Tributyl Phosphate (TBP) ($DN_{SbCl_5} = 23.7$) 151
12. Dimethylformamide (DMF) ($DN_{SbCl_5} \sim 27$) 152
13. N,N-Dimethylacetamide (DMA) ($DN_{SbCl_5} = 27.8$) 154
14. Dimethyl Sulphoxide (DMSO) ($DN_{SbCl_5} = 29.8$) 155
15. Hexamethylphosphoramide (HMPA) ($DN_{SbCl_5} = 38.8$) 159

Chapter VIII

Coordination Chemistry of Certain Transition Metal Ions in Donor Solvents

1. Iodide Ions as Competitive Ligands 161
2. Bromide Ions as Competitive Ligands 162
3. Chloride Ions as Competitive Ligands 163
4. Azide Ions as Competitive Ligands 166
5. Thiocyanate Ions as Competitive Ligands 167
6. Cyanide Ions as Competitive Ligands 167
7. An Attempt to Assign a Donor Number to Anions 168
8. Conclusion 168

Index 169

,, Corpora non agunt nisi fluida"

Chapter I

General

1. Introduction

It was recognized long before the advent of science, that reactions were best carried out in the liquid phase. Whilst with the development of organic chemistry various solvents such as alcohols, ethers or aromatic hydrocarbons have been found to serve as convenient media for numerous types of reactions, inorganic chemistry was mainly developed in solutions in water. As a consequence analytical chemistry, physical chemistry and electrochemistry were also mainly concerned with the study of aqueous systems.

At the turn of the last century liquid ammonia was successfully introduced into inorganic chemistry and was later followed by various other solvents. It is certainly true that the present development of inorganic chemistry would hardly be possible without the extensive use of non-aqueous solvents.

Non-aqueous solvents have been found among different classes of compounds. Each solvent has its characteristic properties, but it is difficult to give a fair account of its character simply by the sum of its physical and chemical properties.

Chemistry is still an experimental science. The choice of a proper solvent for a particular chemical reaction still rests on the experience of the particular research worker. The choice clearly depends upon the reactants used and upon the compound that is being prepared, but no reliable system is available showing relations between certain (and measurable) solvent properties and the characteristics of the reactions or of the products desired. Coordination chemistry has offered a very important approach to different problems in chemistry, and a number of examples illustrating this point are included later in the text.

2. Classification of Solvents

Non-aqueous solvents in a broad sense may be defined as media other than water which will dissolve a reasonable number of compounds and allow the occurrance of chemical reactions. Non-aqueous solvents are found in different classes of chemical compounds which may be classified according to the characteristic features of chemical bonding occurring in them[1].

Those which are most widely used are the "molecular liquids"; they are liquid at or near room temperature and it is of solvents from this class which are usually considered as "Non-Aqueous-Solvents" and which will exclusively be discussed in this presentation.

Molten salt media are the largest class of non-aqueous inorganic solvents, and are known as excellent media for preparations of various compounds. They consist of ionic melts in the liquid state and cover a wide temperature range.

The third class of non-aqueous solvents consists of low melting metals in their liquid states, such as liquid sodium or liquid mercury.

The wide scope of non-aqueous solvent chemistry is well demonstrated by Fig. 1, which clearly shows that only very limited ranges of the vast field of solvents and mixtures of solvents have ever been investigated.

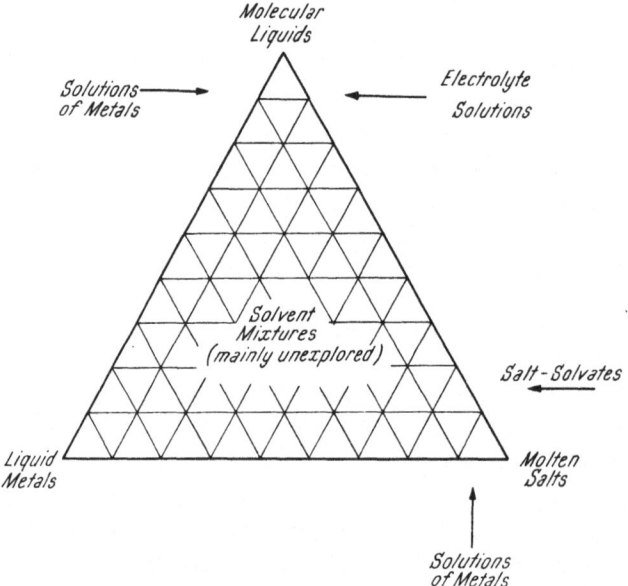

Fig. 1. Types of Solvents

A good deal of solution chemistry is centered around the corners (mainly around the top corner) of the triangle in Fig. 1, whilst the vast field of mixtures of different types of solvents is practically unexplored.

The following account is designed to give an outline of coordination chemistry in molecular liquids as solvents and will cover certain aspects of solution chemistry on the top corner of the triangle.

Molecular liquids as solvents are frequently classified as "Ionizing Solvents" and "Non-Ionizing Solvents"[2]. Ionizing solvents are described as being inherently

---

[1] ADDISON, C. C.: presented at the Int. Conf. Non-Aqueous Solvent Chemistry, McMaster Univ., Hamilton/Ontario, 1967.

[2] AUDRIETH, L. F., and F. KLEINBERG: "Non-Aqueous Solvents," J. Wiley and Sons, New York 1953. — V. GUTMANN, Quart. Revs. (London) 10, 451 (1956).

polar and will dissolve many ionic and covalent compounds to give conducting solutions[2] with a marked tendency to solvate solutes or ions; this is frequently demonstrated by the formation of crystalline compounds containing solvent molecules (solvates). A sharp distinction between ionizing and non-ionizing solvents is as impossible as a distinction between a conductor of an electric current and an insulator. Thus a covalent liquid such as stannic chloride[3] may or may not be called an ionizing solvent, as it dissolves certain covalent compounds to give solutions of very low conductivities. Since interactions between solvent molecules and solute are ineritable it will be hard to define any solvent as "inert". Benzene is in many cases far from "inert", as can be seen from its interactions with iodine or from its ionized solutions of silver perchlorate.

The suggestion was made to classify solvents according to those which are "water-like" and those which are not, but no unambiguous definitions could be given[4].

With the development of the definitions of acids and bases due to BRÖNSTED[5] and LOWRY[6] many solvents were found to which these definitions could be applied. It followed, therefore, that solvents were classified as "proton containing solvents" and "proton-free solvents"[7]. The former contain hydrogen in an ionizable form and the cations produced by auto-ionization are considered to be solvated protons. Typical proton-containing solvents are: water, liquid ammonia, hydrazine, hydrogen cyanide, sulphuric acid, nitric acid, acetic and formic acids, alcohols and amides. Liquid hydrogen fluoride and the other hydrogen halides may also be included in this group of solvents.

From a structural point of view the solvents may be distinguished according to their ability to form hydrogen bonds in the liquid state. Thus water, and other proton-containing solvents may be considered as hydrogen bonded systems in contrast to proton-free solvents, which are either not associated or associated by other structural features, such as oxide- or halide-bridging.

A clear distinction is not possible, since some of them such as nitriles or dimethyl sulphoxide have certain tendencies to ionize a proton from a CH_3-group:

$$A + CH_3CN \rightleftharpoons AH^+ + CH_2CN^-$$
$$A + CH_3SOCH_3 \rightleftharpoons AH^+ + CH_3SOCH_2^-$$

This is known to occur only in the presence of strong proton acceptors and is not taking place to any significant degree in the pure liquid solvents.

From the point of coordination chemistry a distinction between coordinating and non-coordinating solvents was suggested[8]. A true non-coordinating solvent will hardly serve its purpose as a solvent for a reasonable number of chemical compounds or reactions.

[3] SPANDAU, H., and H. HATTWIG: Z. anorg. allg. Chem. **295**, 281 (1958).
[4] JANDER, G.: „Die Chemie in wasserähnlichen Lösungsmitteln", Springer-Verlag, Berlin-Göttingen-Heidelberg 1949.
[5] BRÖNSTED, J. N.: Rec. Trav. Chim. Pays-Bas **42**, 78 (1923), J. physic. Chem. **20**, 777 (1926), Ber. **61**, 2079 (1928), Z. physik. Chem. (A) **163**, 307 (1933), **162**, 128 (1933), **169**, 52 (1937).
[6] LOWRY, T. M.: J. Soc. Chem. Ind. **42**, 43 (1923).
[7] SPANDAU, H., and V. GUTMANN: Angew. Chem. **64**, 93 (1952).
[8] DRAGO, R. S., and K. F. PURCELL in: "Non-Aqueous Solvent Systems." Ed. T. C. WADDINGTON, Academic Press, London and New York 1965.

According to the coordination chemistry approach solvents may be divided into

1. Donor solvents, which have a high tendency to react with electron pair acceptors, and

2. Acceptor solvents, which tend to react with electron pair donors.

Many characteristic features will be different for these two classes of solvents. The former will solvate in the first place metal cations, while the latter will solvate mainly anions. For the consideration of acid-base phenomena in the sense of BRÖNSTED, and of association phenomena through hydrogen bridges, each of these groups of solvents may again be divided into

a) Proton-containing or hydrogen bridged solvents and

b) Proton-free solvents or not hydrogen bridged solvents.

Molten salts and liquid metals are considered to be special classes of non protonic solvents, which will not be discussed in the present treatment.

3. Physical Properties

Both the physical properties and the general chemical character determine the properties of a solvent. Again a clear distinction between them cannot be made and it is hard to estimate the contribution of a particular property to the general solvent properties. The usefulness of a solvent is also influenced by certain other factors, such as convenient liquid range, the ease of purification and handling, or its physiological properties.

Some of the numerous physical properties which to varying extents contribute to the overall solvent character, may be summarized: melting point, boiling point, vapor pressure, refractive index, density, viscosity, heat of vaporization, surface tension, dipole moment and the dielectric constant. The specific conductivity may be considered as a criterion of solvent purity and may in part be due to the presence of impurities. The boiling point may be important in certain displacement reactions, when a solvent of poor donor properties may displace a molecule of higher donor properties by simply disturbing the equilibrium due to its high boiling point. Density and viscosity will be important for the mobilities of ions and for the readiness to facilitate chemical reactions. A high enthalpy of vaporization is an indication of association of solvent molecules in the liquid state; this latter property is frequently represented by the Trouton constant defined by the quotient of the heat of vaporization and the boiling point.

A high dipole moment μ will contribute to the chemical interaction between a polar solvent and ions as well as between the solvent molecules themselves and will, although physical in nature, be discussed in relation to the chemical properties of the solvent. According to the concept of soft and hard acids and bases[9] the degree of polarizability is also important and will be discussed later.

It has been recognized long ago that a reasonably high dielectric constant is an important requirement for an ionizing solvent. With increasing dielectric constant the charged particles in solution will be kept apart and will find it more difficult to form ion pairs or higher aggregated ionic species, which will not

---

[9] PEARSON, R. G.: J. Amer. Chem. Soc. **85**, 3533 (1963).

contribute to the conductivity of the solution. Thus conductivity data cannot be interpreted properly unless the dielectric constant of the solvent (or more precisely that of the solution) is taken into consideration. The DEBYE-HÜCKEL theory and its further development due to ONSAGER and later to FUOSS and KRAUSS led to an understanding of solutions of low conductivity and hence to that of solutions of low dielectric constant. It is also known that the value of the dielectric constant of the medium is contributing to the rates and the mechanisms of various types of reactions. On the other hand the role of the dielectric constant has frequently been overrated. Some new text books make statements such as: "a low dielectric constant results in a general decrease in solubility of a salt"[10] or "the solvation energy increases with increasing dielectric constant of a solvent"[11]. Such statements are confusing the issue since it would remain unexplained, why perchloric acid is nearly undissociated in solutions of anhydrous sulphuric acid ($\varepsilon = 85$), while it is strongly dissociated in anhydrous acetic acid ($\varepsilon = 7$), and why anhydrous hydrogen cyanide with its high dielectric constant of 123 is a poor solvent while tributylphosphate or pyridine with their relatively small dielectric constants of 6.8 and 12.3 respectively are extremely useful solvents for a number of ionic compounds[12]. In fact ionization as distinct from dissociation does not appear to be determined by the dielectric constant[13] as will be described later.

4. Acids and Bases (Acidic and Basic Function)

A considerable amount of coordination chemistry in non-aqueous solutions is frequently regarded as acid-base-type reactions. Numerous solvents are described as acids or bases respectively, although some confusion exists about the meaning of the terms "acid" or "base" due to the existence of different definitions. Before discussing them it should be emphasized that a substance can never be inherently an acid or base respectively; it can, however, be made to act as either of them according to the conditions, so that one should rather use the terms "acidic function" and "basic function."

A. Classical Definition

Before systematic chemistry was developed an acid was described as a substance which may change the color of certain dye-stuffs, has a sour taste and can dissolve metals, but which loses such properties on contact with bases.

S. ARRHENIUS, the father of the ionic theory, postulated the self-dissociation of water and defined an acid as a substance which increases the hydrogen ion

[10] DAY, M. C. Jr., and F. SELBIN: "Theoretical Inorganic Chemistry", p. 340, Reinhold Publ. Corp., New York 1962.

[11] KLEINBERG, F., W. F. ARGERSINGER, and E. GRISWOLD: "Inorganic Chemistry", Heath and Co., Boston 1960, p. 161.

[12] GUTMANN, V.: Coord. Chem. Revs. 2, 239 (1967).

[13] GUTMANN, V., and E. WYCHERA: Rev. Chim. Min. 3, 941 (1966).

concentration and a base as a substance which increases the concentration of hydroxyl ions in the solution. These definitions are clearly restricted to aqueous solutions.

B. The Protonic Concept

A more general definition was suggested independently by BRÖNSTED[5] and LOWRY[6]. Proton-transfer reactions are considered as responsible for both the self-ionization of the amphoteric solvent molecules and for most acid-base reactions in their solutions. Acids and bases are defined as proton donors and proton acceptors respectively and acid-base reactions are regarded as being due to proton transfer reactions (protolysis). The most significant difference from the ARRHE-NIUS definition is that the proton itself is neither acid nor base. Even the solvent molecules can act either as acids or bases, a phenomenon which is responsible for the autoprotolysis of the pure liquid solvents.

Typical reactions in water may be represented according to:

$$H^+$$

$$
\begin{array}{llll}
\text{Autoprotolysis:} & H_2O + H_2O \rightleftharpoons OH^- + H_3O^+ \\
\text{Acidic Solute:} & HCl + H_2O \rightleftharpoons Cl^- + H_3O^+ \\
\text{Basic Solute:} & H_2O + NH_3 \rightleftharpoons OH^- + NH_4^+ \\
& \text{Acid I} \quad \text{Base II} \quad \text{Base I} \quad \text{Acid II}
\end{array}
$$

Each acid is converted into a corresponding base and *vice versa*. If the acid is weak (water) the corresponding base is strong (OH^-), if the acid is strong (HCl), the corresponding base (Cl^-) is weak. It is clear that the acidic function is not restricted to neutral compounds as the NH_4^+ ion is a weak, and the H_3O^+ ion a strong acid in water, while the amphoteric HSO_4^- ion can act either as an anion-base by accepting a proton or as an anion-acid by donating a proton:

$$H^+$$

$$HSO_4^- + H_3O^+ \rightleftharpoons H_2SO_4 + H_2O$$
base

$$H^+$$

$$HSO_4^- \quad H_2O \rightleftharpoons SO_4^{--} + H_3O^+$$
acid

The activity of the hydrated protons is a quantitative measure of the acidity of the solution and it can be measured by electrochemical methods with a high degree of accuracy. Another advantage is the fact that ionization of a compound, neutralization and hydrolysis (or more general solvolysis) are all considered as protolytic reactions. When K_w is the ionic product of water or of the protonic solvent used and K_a the dissociation constant of the acid,

$$K_{solvolysis} = \frac{K_w}{K_a}$$

This concept is clearly not restricted to aqueous systems, but can also be applied to protolytic reactions in other systems, such as in solutions of liquid ammonia, hydrogen cyanide, sulphuric acid, alcohols, as well as even in benzene (when a protonic acid is reacted with a base in this medium). It is, however, not applicable to reactions in which protons are not involved.

The asymmetry in the definition has been critisized[14]. The acidic function is limited to a compound or ion that contains hydrogen in an ionizable form, whilst any substance or ion capable of combining with a proton is exerting a basic function.

C. The Solvent-System Concept

GERMANN[15] found that aluminium trichloride reacts with carbonyl chloride to give a solution which yields $KAlCl_4$ on addition of potassium chloride. He considered aluminium chloride as a solvo-acid and potassium chloride as a solvo-base, according to the following self-ionization equilibrium of the pure solvent (which is unlikely to be correct):

$$COCl_2 \rightleftharpoons CO^{++} + 2\ Cl^- \qquad \text{(self ionization)}$$
$$COCl_2 + AlCl_3 \rightleftharpoons CO^{++} + 2\ AlCl_4^-$$

CADY and ELSEY[16] gave the so-called "solvent-system definitions," which depend on the mode of self-ionization of the particular solvent. Each solvent is considered as a parent of acids and bases. A solvo-acid is defined as a solute which increases the concentration of cations characteristic of the pure solvent and a solvo-base as a solute which increases that of the anions characteristic of the pure solvent:

In liquid ammonia[17]:   self-ionization $2\ NH_3 \rightleftharpoons NH_4^+ + NH_2^-$
                     solvo-acid $NH_4Cl \rightleftharpoons NH_4^+ + Cl^-$
                     solvo-base $NaNH_2 \rightleftharpoons Na^+ + NH_2^-$

In BrF_3[18,19]:   self-ionization $2\ BrF_3 \rightleftharpoons BrF_2^+ + BrF_4^-$
                     solvo-acid $SbF_5 \cdot BrF_3 \rightleftharpoons BrF_2^+ + SbF_6^-$
                     solvo-base $KBrF_4 \rightleftharpoons K^+ + BrF_4^-$

In $HgBr_2$[20]:   self-ionization $2\ HgBr_2 \rightleftharpoons HgBr^+ + HgBr_3^-$
                     solvo-acid $HgSO_4 + HgBr_2 \rightleftharpoons 2\ HgBr^+ + SO_4^{--}$
                     solvo-base $KBr + HgBr_2 \rightleftharpoons K^+ + HgBr_3^-$

It has recently been pointed out that the solvent system concept based on the mode of the self-ionization of a solvent has been overemphasized[8], but it has definitely served a useful purpose in the exploration of reactions in a number of non-aqueous solutions.

[14] EBERT, L., and N. KONOPIK: Oest. Chem. Ztg. 50, 184 (1949).

[15] GERMANN, A. F. O.: J. Amer. Chem. Soc. 47, 2461 (1925).

[16] CADY, H. P., and H. M. ELSEY: J. Chem. Educ. 5, 1425 (1928).

[17] FRANKLIN, E. C.: "The Nitrogen System of Compounds," Reinhold Publ. Corp., New York 1935.

[18] SHARPE, A. G., and H. J. EMELÉUS: J. Chem. Soc. 1948, 2135.

[19] BANKS, A. A., H. J. EMELÉUS, and A. A. WOOLF: J. Chem. Soc. 1949, 2861.

[20] JANDER, G., and K. BRODERSEN: Z. anorg. allg. Chem. 261, 261 (1950); 262, 33 (1950).

Whenever these definitions are used the terms "solvo-acid" and "solvo-base" should be applied in order to distinguish from acids and bases in the sense of the theory by BRÖNSTED.

JANDER has made use of the solvent-system conceptions, but preferred to use the term "acid-analogon" in the place of solvo-acid and "base-analogon" in the place of solvo-base to stress the analogy with the reactions in water[4].

The solvent-system concept definitions are, however, restricted to solvents with characteristic self-ionization.

D. The Ionotropic Definitions

LUX described reactions in oxide melts in terms of oxide ion transfer reactions (oxidotropism)[21, 22]. A base is defined as an oxide ion donor and an acid as an oxide ion acceptor:

$$
\begin{array}{cccc}
O^{--} & & O^{--} & \\
\lfloor\qquad\downarrow & & \downarrow\qquad\rceil & \\
CaO + & CO_2 \rightleftharpoons & Ca^{++} + & CO_3{}^{--} \\
\text{Base I} & \text{Acid II} & \text{Acid I} & \text{Base II}
\end{array}
$$

Donor- and acceptor functions for base and acid respectively are now opposite to those for BRÖNSTED-systems, since the ion responsible for the reactions (the oxide ion) is carrying negative charges.

EBERT and KONOPIK[14] distinguish between donor and acceptor functions for both acids and bases. A donor acid releases solvent cations while an acceptor acid accepts solvent anions e.g. in water[14].

donor acid: $HCl + H_2O \rightleftharpoons H_3O^+ + Cl^-$
acceptor acid: $CO_2 + OH^- \rightleftharpoons HCO_3^-$

Likewise a donor base releases solvent anions and an acceptor base accepts solvent cations, e.g. in water:

donor base: $NaOH \rightleftharpoons Na^+ + OH^-$
acceptor base: $NH_3 + H^+ \rightleftharpoons NH_4^+$

By making use of these considerations GUTMANN and LINDQVIST[23] extended the BRÖNSTED theory by replacing the term "prototropic reactions" by "ionotropic reactions". In general solvent systems are classified as
   a) cationotropic,
   b) anionotropic.

The former include the BRÖNSTED theory for all prototropic systems, while the latter may be oxidotropic, fluoridotropic, chloridotropic etc. For certain systems such as liquid hydrogen fluoride both the protonic and the fluoridotropic approaches are possible[24], since the formation of ions from the associated solvent molecules may be considered as due to either proton- or fluoride ion-transfer[23, 24].

[21] LUX, H.: Z. Elektrochem. **45**, 303 (1939); **52**, 220 (1948); **53**, 41, 43, 45 (1949).
[22] LUX, H.: Z. anorg. allg. Chem. **250**, 159 (1942).
[23] GUTMANN, V., and I. LINDQVIST: Z. physik. Chem. **203**, 250 (1954).
[24] GUTMANN, V.: Svensk Kem. Tidskr. **68**, 1 (1955).

$$H^+$$

$$(\overbrace{HF})_{n+1} + (\overbrace{HF})_m \rightleftharpoons [(HF)_nF]^- + [(HF)_mH]^+$$

$$F^-$$

$$(\underbrace{HF})_n + (\underbrace{HF})_{m+1} \rightleftharpoons [(HF)_nF]^- + [(HF)_mH]^+$$

For covalent fluorides as solvents fluoride ion-transfer is postulated and for covalent chlorides chloride ion-transfer:

$$F^-$$

$$BrF_3 + BrF_3 \rightleftharpoons BrF_2^+ + BrF_4^-$$

$$Cl^-$$

$$AsCl_3 + AsCl_3 \rightleftharpoons AsCl_2^+ + AsCl_4^-$$

Indeed fluorine and chlorine bridges seem to be responsible for the association in the pure liquid states, in analogy to hydrogen bridging occuring in many protonic solvents.

When the attempt is made to reach definitions independent of a solvent they become similar to those given by USSANOVICH[25] according to which an acid is either a cation donor or an anion acceptor, and a base either an anion donor or a cation acceptor. In this way certain redox reactions are also considered as acid-base reactions.

E. The Lewis Concept

LEWIS defined in a very general way an acid as an electron pair-acceptor and a base as a electron-pair donor[26]. The criterion for the occurrance of an acid-base reaction is the formation of a coordinate covalent link[26, 27], e.g.

$$(CH_3)_3N: + BF_3 \rightleftharpoons (CH_3)_3N \rightarrow BF_3$$

In this way both the acidic and the basic function is independent of an ion-transfer mechanism and independent from a solvent. In fact it covers all reactions considered in coordination chemistry. The terms "Lewis acid" and "Lewis base" are applied mainly in the field of organic chemistry.

In order to avoid confusion with acids in the BRÖNSTED sense, coordination chemists may prefer the terms "acceptor" and "donor" in the place of Lewis acid and Lewis base respectively and we shall follow this convention in the course of this presentation.

KOLTHOFF[28] suggested the term "Protoacid" for Lewis acids except proton donors, and BJERRUM[29] prefers the term "Antibase."

[25] USSANOVICH, M.: Zh. obshch. Khim. USSR **9**, 182 (1939).

[26] LEWIS, G. N.: J. Franklin Inst. **226**, 293 (1938).

[27] LUDER, W. F., and S. ZUFFANTI: "The Electronic Theory of Acids and Bases," Wiley, New York 1946.

[28] KOLTHOFF, I. M.: J. physic. Chem. **48**, 51 (1944).

[29] BJERRUM, J.: Angew. Chem. **63**, 527 (1951)

F. Hard and Soft Acids and Bases

For the special case of metal ions as Lewis acids AHRLAND, CHATT and DA-VIES[30] made a useful classification. The metal ions were divided into two classes depending on whether they formed their most stable complexes with the first ligand atoms of each group in the periodic table class (a) or whether they formed their most stable complexes with the second or a subsequent member of each group, class (b)[31].

(a) $N \gg P > As > Sb$
(b) $N \ll P > As > Sb$
(a) $O \gg S > Se > Te$
(b) $O \ll S \sim Se \sim Te$
(a) $F \gg Cl > Br > I$
(b) $F \ll Cl < Br < I$

The proton is the most typical class (a) ion. PEARSON[9] termed Lewis acids of class (a) character as "hard" acids and those of class (b) character as "soft" acids. In classifying Lewis acids the criterion of AHRLAND, CHATT and DAVIES[30] was used whenever possible, but other rules were introduced. "Soft" acids are expected to combine with "soft" bases and "hard" acids prefer to combine with "hard" acids. "Soft" bases are polarizable (high polarizability) and "hard" bases are non-polarizable.

Table 1. *Classification of Electronpair Acceptors (Lewis Acids)*

Hard	Soft
H^+, Li^+, Na^+, K^+	Cu^+, Ag^+, Au^+, Tl^+, Hg^+
Be^{2+}, Mg^{2+}, Ca^{2+}, Sr^{2+}, Mn^{2+}	Pd^{2+}, Cd^{2+}, Pt^{2+}, Hg^{2+}, CH_3Hg^+,
	$Co(CN)_5^{2-}$, Pt^{4+}, Te^{4+}
Al^{3+}, Sc^{3+}, Ga^{3+}, In^{3+}, La^{3+}, N^{3+}, Cl^{3+},	Tl^{3+}, $Tl(CH_3)_3$, BH_3, $Ga(CH_3)_3$, $GaCl_3$,
Gd^{3+}, Lu^{3+}	GaI_3, $InCl_3$
Cr^{3+}, Co^{3+}, Fe^{3+}, As^{3+}	RS^+, RSe^+, RTe^+
Si^{4+}, Ti^{4+}, Zr^{4+}, Th^{4+}, U^{4+}, Pu^{4+}, Ce^{3+}, Hf^{4+}	I^+, Br^+, HO^+, RO^+
UO_2^{2+}, $(CH_3)_2Sn^{2+}$, VO^{2+}, MoO^{3+}	I_2, Br_2, ICN, etc.
$Be(CH_3)_2$, BF_3, $B(OR)_3$	trinitrobenzene etc.
$Al(CH_3)_3$, $AlCl_3$, AlH_3	chloranil, quinones, etc.
RPO_2^+, $ROPO_2^+$	tetracyanoethylene, etc.
RSO_2^+, $ROSO_2^+$, SO_3	O, Cl, Br, I, N
I^{7+}, I^{5+}, Cl^{7+}, Cr^{6+}	M^0 (metal atoms)
RCO^+, CO_2, NC^+	
HX (hydrogen bonding molecules)	carbenes

Borderline

Fe^{2+}, Co^{2+}, Ni^{2+}, Cu^{2+}, Zn^{2+}, Pb^{2+}, Sn^{2+}, Sb^{3+}, Bi^{3+}, Rh^{3+}, Ir^{3+}, $B(CH_3)_3$, SO_2, NO^+, Ru^{2+}, As^{3+}, R_3C^+, $C_6H_5^+$, GaH_3.

The polarizability of ions and molecules was in fact introduced by FAJANS[32] as early as in 1923, but it is understood that various other factors contribute to "hard" and "soft" properties.

[30] AHRLAND, S., J. CHATT, and N. R. DAVIES: Quart. Revs. (London) **12**, 265 (1958).

[31] AHRLAND, S.: "Structure and Bonding," Ed. C. K. JØRGENSEN, Springer-Verlag, Berlin-Heidelberg-New York, Vol. I, 207 (1966).

[32] FAJANS, K.: Chem. Eng. News **43**, [22], 96 (1965).

5. Techniques with Non-Aqueous Solvents

Since many non-aqueous solvents are reactive, corrosive, toxic or highly volatile or have a combination of these undesirable properties, the experimental difficulties may be considerable. The reactivity of the solvents may require the exclusion of oxygen or carbon dioxide and most likely that of water. Since it is extremely difficult and frequently impossible to remove the last traces of water or to prevent its admission in trace quantities, the terms "non aqueous" and "anhydrous" should be understood in the sense of "extremely low water content."

Each solvent will therefore require special techniques in purification and handling and reference can be made to the excellent presentation by POPOV[33].

[33] POPOV, A. I.: Chapter 2 in Vol. I, "Techniques in Inorganic Chemistry," Ed. H. B. JONASSEN, and A. WEISSBERGER, p. 37—102, Interscience, New York, London 1963.

Chapter II

Principles of Coordination Chemistry in Non-Aqueous Solutions

1. Donor Solvents and Acceptor Solvents

According to their different chemical properties solvents may be divided into donor solvents and acceptor solvents.

A donor solvent (D) will in general react with acceptor molecules and acceptor ions:

$$SbCl_5 + D \rightleftharpoons D \cdot SbCl_5$$
$$Co^{++} + 6\,D \rightleftharpoons [CoD_6]^{++}$$

Since most metal ions are electron pair acceptors interaction will take place between metal ions and donor solvent molecules leading to solvated metal cations.

An acceptor solvent (A) will in general react with donor compounds. Since most anions are electron pair donors, solvation of anions will occur by interaction with the solvent molecules, e.g.:

$$(C_6H_5)_3CCl + A \rightleftharpoons (C_6H_5)_3C^+ + ACl^-$$
$$KF + A \rightleftharpoons K^+ + AF^-$$

This type of interaction is usually taking place to a smaller degree than solvation of a cation by a donor solvent.

Some of the solvents will be able to act both as weak acceptor or as weak donor solvents, for example hydrogen fluoride, and it is therefore difficult to assign them to any of these groups.

2. Coordinating Properties of Solvents

A. Donor Strength

The first step in the process of dissolution is the chemical interaction between solute and solvent molecules. When the solute has a molecular lattice, little energy is required to overcome the VAN DER WAAL's forces and a fairly inert solvent will usually be capable of dissolving such compounds. With higher lattice energy of the solute more energy is required to disrupt the crystal lattice. This energy is provided by the chemical reaction either between solute A and the donor solvent D or between acceptor solvent A and solute D[1, 2].

$$A + nD \to AD_n; \quad -\Delta H_{AD_n}$$
solvent
$$nA + D \to A_nD; \quad -\Delta H_{A_nD}$$
solvent

If ΔH is higher than the lattice energy the process of dissolution will be exothermic, if it is smaller, the reaction will be endothermic and if ΔH is much smaller than the lattice energy dissolution will scarcely occur[1, 2].

Most molecules or ions in solution are solvated. The anhydrous Cu^{++} is colourless as is exemplified in anhydrous copper sulphate. The hydrated ion $[Cu(OH_2)_4]^{2+}$ is blue in color and the ammoniated ion $[Cu(NH_3)_4]^{2+}$ is deep blue. BRÖNSTED was well aware of the importance of solvent coordination when he formulated the acidic properties of metal ions. They can only be explained by proton donation of the hydrated (solvent-coordinated) ion:

$$[Fe(OH_2)_6]^{3+} \rightleftharpoons [Fe(OH)(OH_2)_5]^{2+} + H^+$$

Thus the "Coordination Model of Non-Aqueous Solutions" suggested by DRAGO[3, 4] is certainly not new.

The majority of useful solvents have donor properties, and the ligands to be coordinated to an acceptor molecule or to an acceptor ion will have to compete for coordination with solvent molecules. Such reactions in solution may be represented as replacement reactions of solvent molecules coordinated to ions or molecules by competitive ligands L or X^- which may be neutral or charged. The occurrance of replacement or ligand exchange reactions will depend on the relative donor properties of solvent molecules D and competitive ligands L or X^- towards the ion or molecule under consideration:

$$AD + L \rightleftharpoons AL + D$$
$$AD + X^- \rightleftharpoons AX^- + D$$

Thus coordination reactions including those of solvent-coordination or solvation can be regarded as donor-acceptor reactions, for which the donor strength of the solvent molecules will be a decisive factor[5].

BRIEGLEB[6] pointed out that the "donor-strength" as an absolute property of a molecule is represented by its ionization energy. The donor strength may alternatively be determined relative to a certain reference acceptor: ΔH- and ΔG-values for the reactions of various donors with a reference acceptor may be obtained in an "inert" solvent[7]. The reference acceptor must react with the donor-molecules in a well defined way and should preferably form 1 : 1 compounds with

[1] GUTMANN, V.: Coord. Chem. Revs., **2**, 239 (1967).

[2] GUTMANN, V.: "Emeléus Volume," to be published by Cambridge University press.

[3] DRAGO, R. S., and D. W. MEEK: J. Phys. Chem. **65**, 1446 (1961).

[4] MEEK, D. W., and R. S. DRAGO: J. Amer. Chem. Soc. **83**, 4322 (1961).

[5] GUTMANN, V., and E. WYCHERA: Rev. Chim. Min. **3**, 941 (1967).

[6] BRIEGLEB, G.: „Elektronen-Donator-Akzeptor-Komplexe", Springer-Verlag, Berlin-Göttingen-Heidelberg 1961.

[7] LINDQVIST, I., and M. ZACKRISSON: Acta Chem. Scand. **14**, 453 (1960).

the donor molecules; the reaction should be practically complete to obtain accurate results[5].

Until recently only qualitative information was available concerning the donor strength of a solvent. LINDQVIST and ZACKRISSON[7] reported a scale of relative donor strengths for various donor molecules—including some which are in use as non-aqueous solvents—as a result of comparative calorimetric measurements with $SbCl_5$ or $SnCl_4$ as reference acceptors. They gave the following order[7]:

$(C_6H_5)_2SeO \sim (C_6H_5)_3AsO > (CH_3O)_3PO > (CH_3)_2SO \gg (C_2H_5)_2S \sim (CH_3)_2CO \sim$
$CH_3COOC_2H_5 \sim (C_2H_5)_2CO \sim (C_2H_5)_2O \gg (CH_3)_2SO_2 > (C_6H_5)_2SO_2 \sim POCl_3 \sim$
$SeOCl_2 > SOCl_2$

and concluded that the order of donor strength of a M=O group, such as Se=O or P=O is decreased by substituents at M in the following order[7, 8]:

$$RO \geq R > C_6H_5 > Cl^-$$

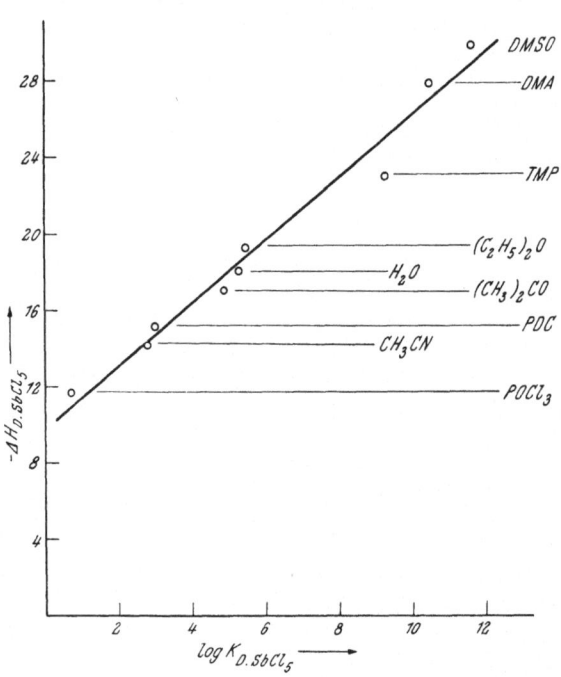

Fig. 2. Relation between $-\Delta H_{D \cdot SbCl_5}$ and $\log K_{D \cdot SbCl_5}$ for different donor solvents D

It is not surprising that no further progress was made as long as no quantitative data was produced. In a recent article the statement is found[9]: "triethylphosphate is expected to be a slightly better donor than phosphorus oxychloride, but the similarity is such, that vastly different coordination behaviour would not be expected." The authors[9] went as far as to question the interpretation

[8] BAAZ, M., and V. GUTMANN in: "Friedel-Crafts and Related Reactions," Ed. G. OLAH, Vol. I, p. 367, Interscience, New York 1963.
[9] DRAGO, R. S., and K. F. PURCELL: Chapter 5 in "Non-Aqueous Solvent Systems", Ed. T. C. WADDINGTON, Acad. Press, London, New York 1965.

of results in phosphorus oxychloride simply because different conclusions were obtained in the system ferric chloride-triethylphosphate[4].

It is now apparent that the differences in donor properties of the two solvents are such that different behaviour of the solutions is to be expected corroborating what is actually found to occur.

Quantitative calorimetric measurements on the interactions of a number of O- and certain N-containing solvent molecules with antimony(V) chloride as reference acceptor in dichlorethane are now available[10]:

$$D_{(dissolved)} + SbCl_{5(dissolved)} \rightleftharpoons D \cdot SbCl_{5(dissolved)}; \quad -\Delta H_{D \cdot SbCl_5}$$

By using highly dilute solutions the conditions of the gas phase reactions were approached.

For the same equilibria the ΔG-values were obtained from spectrophotometric[10] and NMR-measurements[11] and compared with the $-\Delta H_{D \cdot SbCl_5}$-figures (Table 2, Fig. 2).

Table 2. $-\Delta H_{D \cdot SbCl_5}$ and $\log K_{D \cdot SbCl_5}$-values for Several Donor Solvents D

Donor	$-\Delta H_{D \cdot SbCl_5}$ [kcal $\cdot$ mole^{-1}]	$\log K_{D \cdot SbCl_5}$
Thionyl chloride	0.4	0.3
Acetic Anhydride	10.5	—
Phosphorus Oxychloride	11.7	0.7
Acetonitrile	14.1	2.8
Sulpholane	14.3	—
Propandiol-1,2-carbonate ...	15.1	3.0
Acetone	17.0	4.9
Water	~18.0	5.3
Diethylether	19.2	5.5
Tetrahydrofurane	20.0	—
Trimethylphosphate	23.0	9.3
Tributylphosphate	24.5	—
Dimethylacetamide	27,8	10.5
Dimethylsulphoxide	29.8	11.1

The linear relationship found between $-\Delta H_{D \cdot SbCl_5}$ and $\log K_{D \cdot SbCl_5}$ ($K_{D \cdot SbCl_5}$ being the formation constants of $D \cdot SbCl_5$ with one of the respective solvent molecules D) shows that the entropic contributions are equal for all solvent-acceptor reactions under consideration[10, 11]. It is therefore justified from the thermodynamic point of view to consider the $-\Delta H_{D \cdot SbCl_5}$ values as representative expressions for the degrees of interaction between D and $SbCl_5$[1, 2, 5, 12]. This is confirmed by the analogous relationship observed for the interactions of various D with $(CH_3)_3SnCl$[13].

In order to study the influence of the nature of the acceptor molecules the $-\Delta H_{D \cdot SbCl_5}$ values were compared with the $-\Delta H_{D \cdot A}$ values with A = anti-

[10] GUTMANN, V., A. STEININGER, and E. WYCHERA: Mh. Chem. **97**, 460 (1966).

[11] GUTMANN, V., E. WYCHERA, and F. MAIRINGER: Mh. Chem. **97**, 1265 (1966).

[12] GUTMANN, V.: "Halogen Chemistry", Ed. V. GUTMANN, Vol. II, p. 399 ff. Acad. Press, London and New York 1967.

[13] BOLLES, T. F., and R. S. DRAGO: J. Amer. Chem. Soc. **88**, 3921 (1966).

mony(III)chloride (Fig. 3)[14], antimony(III)bromide (Fig. 4)[14] measured calori-
metrically and with trimethyltinchloride[13] (Fig. 5), phenol (Fig. 6) and iodine
(Fig. 6) calculated from the equilibrium constants between D and A[13,15-25]. It

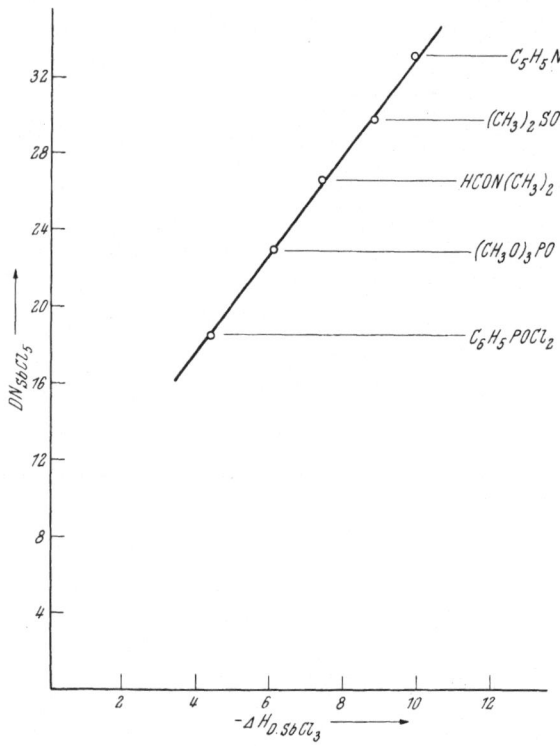

Fig. 3. Relation between DN_{SbCl_5} and $-\varDelta H_{D \cdot SbCl_3}$ for different donor solvents 1)

was found that the figures are proportional to the $-\varDelta H_D \cdot {}_{SbCl_5}$ values (Table 3,
p. 18) showing that specific interactions between the solvent molecules and the
acceptor molecules under consideration are small enough to be neglected.

[14] GUTMANN, V., E. WYCHERA, and S. ANGELOV: Unpublished.

[15] DRAGO, R. S., V. A. MODE, J. G. KAY, and D. L. LYDY: J. Amer. Chem. Soc. **87**, 5010 (1965).

[16] PERSON, W. B., W. C. GOLTON, and A. I. POPOV: J. Amer. Chem. Soc. **85**, 891 (1963).

[17] JOESTEN, M. D., and R. S. DRAGO: J. Amer. Chem. Soc. **84**, 3817 (1962).

[18] MIDDAUGH, R. L., R. S. DRAGO, and R. S. NIEDZIELSKI: J. Amer. Chem. Soc. **86**, 388 (1964).

[19] NAGAKURA, S.: J. Amer. Chem. Soc. **76**, 3070 (1954).

[20] TAMRES, M., and S. M. BRANDON: J. Amer. Chem. Soc. **82**, 2129 (1960).

[21] AKSNES, G., and T. GRAMSTAD: Acta Chem. Scand. **14**, 1485 (1960).

[22] JOESTEN, M. D., and R. S. DRAGO: J. Amer. Chem. Soc. **84**, 2696 (1962).

[23] JOESTEN, M. D., and R. S. DRAGO: J. Amer. Chem. Soc. **84**, 2037 (1962).

[24] DRAGO, R. S., B. B. WAYLAND, and R. L. CARLSON: J. Amer. Chem. Soc. **85**, 3125 (1963).

[25] DRAGO, R. S., D. A. WENZ, and R. L. CARLSON: J. Amer. Chem. Soc. **84**, 1106 (1962).

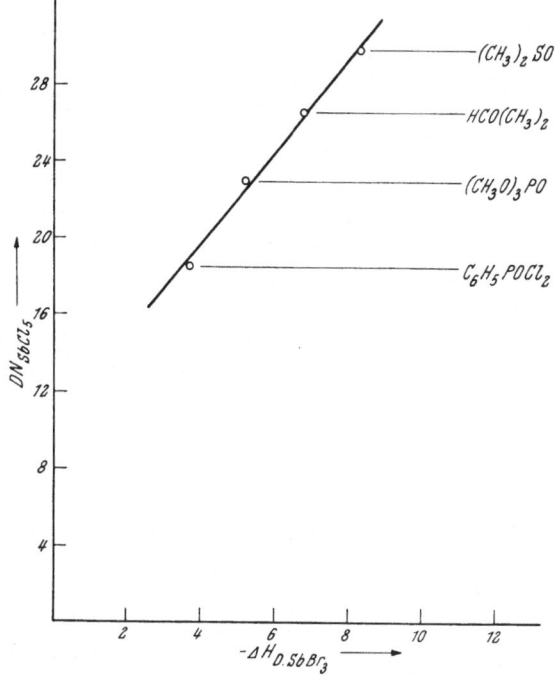

Fig. 4. Relation between DN$_{SbCl_5}$ and $-\Delta H._{DSbBr_3}$ for different donor solvents D

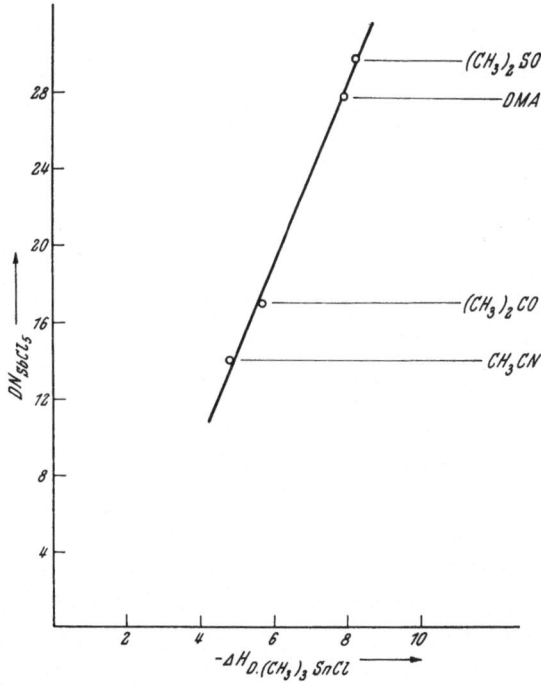

Fig. 5. Relation between DN$_{SbCl_5}$ and $-\Delta H_{D.(CH_3)_3SnCl}$ for different donor solvents D

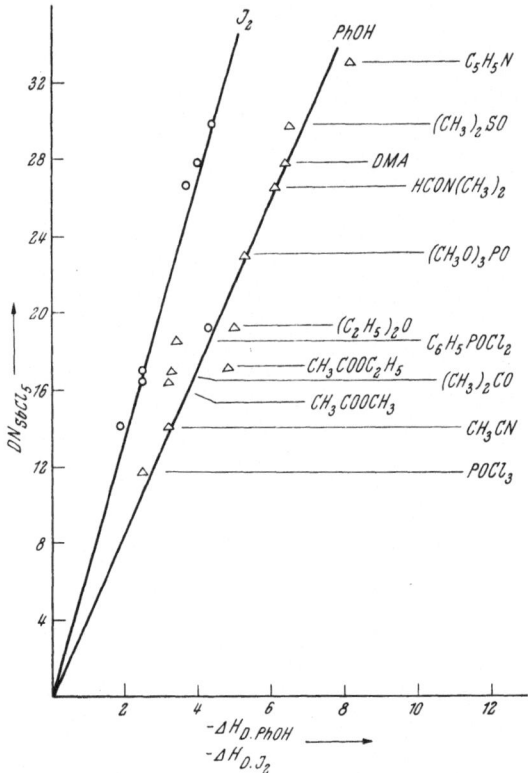

Fig. 6. Relation between DN_{SbCl_5}, $-\Delta H_{D \cdot C_6H_5OH}$ and $-\Delta H_{D \cdot I_2}$ for different donor solvents D

Table 3. $-\Delta H_{D \cdot A}$-values for Reactions of Certain Donor Solvents with Different Acceptor Molecules in "Inert" Media

Donor-Solvent \ Acceptor	$SbCl_5$	$SbCl_3$	$SbBr_3$	C_6H_5OH	I_2	$(CH_3)_3SnCl$
$POCl_3$	11.7	—	—	2.5	—	—
CH_3CN	14.1	—	—	3.2	1.9	4.8
CH_3COOCH_3	16.1	—	—	3.2	2.5	—
$(CH_3)_2CO$	17.0	—	—	3.3	2.5	5.7
$CH_3COOC_2H_5$	17.1	—	—	4.8	—	—
$C_6H_5POCl_2$	18.5	4.4	3.7	3.4	—	—
$(C_2H_5)_2O$	19.2	—	—	5.0	4.3	—
$(CH_3O)_3PO$	23.0	6.1	5.2	5.3	—	—
$HCON(CH_3)_2$	26.6	7.4	6.8	6.1	3.7	—
N-N-Dimethylacetamide .	27.8	—	—	6.4	4.0	7.9
$(CH_3)_2SO$	29.8	8.8	8.3	6.5	4.4	8.2
C_5H_5N	33.1	9.9	—	8.1	7.8	(6.5)

Such relations can only be expected if 1 : 1 reactions are considered, analogous mechanisms are involved (different bonding character in a particular system may lead to deviations, as is known for the systems pyridine-iodine and pyridine-trimethyltinchloride) and if secondary reactions of the complex, such as ionization, do not occur.

B. The Donor Number

These findings lead to the definition of the "donor-number" for each of the solvents. The donor number is defined as the numerical quantity of the $-\Delta H_D \cdot \text{SbCl}_5$ value[26].

$$\text{DN}_{\text{SbCl}_5} \equiv -\Delta H_D \cdot \text{SbCl}_5$$

Table 4. *Donor Number* $\text{DN}_{\text{SbCl}_5}$ *and Dielectric Constants (DEC) of Certain Solvents*

Solvent	$\text{DN}_{\text{SbCl}_5}$	DEC
1,2-Dichloroethane	—	10.1
Sulphuryl chloride	0.1	10.0
Thionyl chloride	0.4	9.2
Acetyl chloride	0.7	15.8
Benzoyl chloride	2.3	23.0
Nitromethane	2.7	35.9
Nitrobenzene	4.4	34.8
Acetic Anhydride	10.5	20.7
Benzonitrile	11.9	25.2
Selenium oxychloride	12.2	46.0
Acetonitrile	14.1	38.0
Sulpholane	14.8	42.0
Propanediol-1,2-carbonate	15.1	69.0
Benzyl cyanide	15.1	18.4
Ethylene sulphite	15.3	41.0
iso-Butyronitrile	15.4	20.4
Propionitrile	16.1	27.7
Ethylene carbonate	16.4	89.1
Phenylphosphonic difluoride	16.4	27.9
Methylacetate	16.5	6.7
n-Butyronitrile	16.6	20.3
Acetone	17.0	20.7
Ethyl acetate	17.1	6.0
Water	18.0	81.0
Phenylphosphonic dichloride	18.5	26.0
Diethylether	19.2	4.3
Tetrahydrofurane	20.0	7.6
Diphenylphosphinic chloride	22.4	—
Trimethyl phosphate	23.0	20.6
Tributyl phosphate	23.7	6.8
Dimethylformamide	26.6	36.1
N,N-Dimethylacetamide	27.8	38.9
Dimethyl sulphoxide	29.8	45.0
N,N-Diethylformamide	30.9	—
N,N-Diethylacetamide	32.2	—
Pyridine	33.1	12.3
Hexamethylphosphoramide	38.8	30.0

The donor number is nearly a molecular property of the solvent, which is easily determined by experiment. It expresses the total amount of interaction with an acceptor molecule, including contributions both by dipole-dipole or dipole-ion

---

[26] GUTMANN, V., and E. WYCHERA: Inorg. Nucl. Chem. Letters **2**, 257 (1966).

interactions and by the binding effect caused by the availability of the free electron pair[2]; to some extent even steric properties of the solvent molecules may be contained in it.

Thus the donor number is considered a semiquantitative measure of solute-solvent interactions. It is recognised that no allowance has been made for specific interactions between certain individual donor-acceptor species. Such refinements are currently impossible owing to a lack of knowledge, and indeed of experimental data, as to the precise nature of such specific interactions.

C. Prediction of Donor-Acceptor Interactions

The approximately linear relationship between the donor number and the $-\Delta H_{A \cdot D}$ values for the interactions between an acceptor A and the donor solvent D allows immediately an empirical approach to predict $-\Delta H_{A \cdot D}$ values[1]. Calorimetric data must be available for the interactions of the given acceptor A with at least two different donors the donor numbers of which must be known. These values may be plotted against the donor numbers of the donors. From this plot $-\Delta H_{A \cdot D}$ for any D with given DN_{SbCl_5} can be readily derived.

DRAGO and WAYLAND[27] have suggested a different approach to predict $-\Delta H_{A \cdot D}$ values according to

$$-\Delta H_{A \cdot D} = C_A \cdot C_D + E_A \cdot E_D$$

when C are the covalent and E the electrostatic contributions denoted by A and D for acceptor and donor respectively. In order to calculate the $-\Delta H_{A \cdot D}$ value such values must be known for the interactions of A with 4 different donors $(D_1, D_2, D_3$ and $D_4)$. With

$$C_{D1} = a \cdot R_{D1}$$
$$E_{D1} = b \cdot \mu_{D1}$$

where R are the polarisabilities and μ the dipole moments, four equations are obtained with four unknown quantities $(E_A, C_A, a$ and b).

When C_A, E_A and the $-\Delta H_{A \cdot D}$ values for two donors are known, calculations of $-\Delta H_{A \cdot D}$ values for other donors can be carried out. Very good agreement between observed and calculated figures is found for A = iodine or phenol[27], but agreement is poor for A = trimethylboron (Table 5) and D = trimethylamine.

The discrepancies are appreciable when $SbCl_5$ is used as an acceptor[1]. The parameters C_{SbCl_5} and E_{SbCl_5} are different for different reference bases. When methylacetate and dimethylformamide are used as reference donors $C_{SbCl_5} = 5.17$ and $E_{SbCl_5} = 2.9$, while with acetone and dimethylsulphoxide as reference donors $C_{SbCl_5} = 2.56$ and $E_{SbCl_5} = 21.70$.

Thus the graphical approach suggested by GUTMANN[1] appears to be more convenient and more reliable than the calculation procedure. Both approaches require the same amount of experimental information, namely two $-\Delta H_{A \cdot D}$ values representing the interactions of the acceptor towards two different donor molecules.

---

[27] DRAGO, R. S., and B. B. WAYLAND: J. Amer. Chem. Soc. **87**, 3751 (1965).

Table 5. *Calculated and Observed* $-\Delta H_{A \cdot D}$-*values for the Reactions of* $A = B(CH_3)_3$ *with Certain N-Bases as Donors*

Donor	$-\Delta H_{A \cdot D}$ (calculated)	$-\Delta H_{A \cdot D}$ (observed)
NH_3		13.75 } used as basis for
CH_3NH_2		17.64 } the calculation
$(CH_3)_2NH$	20.72	19.26
$(CH_3)_3N$	25.82	17.62

Table 6. *Calculations of* $-\Delta H_{D \cdot SbCl_5}$-*values for Different Donors with Methylacetate and Dimethylformamide as Reference Donors*

Donor	$-\Delta H_{D \cdot SbCl_5}$	
	calculated	observed (DN_{SbCl_5})
Acetonitrile	16.03	14.1
Acetone	12.52	17.0
Ethyl acetate	20.75	17.1
Diethylether	26.82	19.2
Dimethylacetamide	28.41	27.8
Tetramethylurea	28.90	29.6
Dimethyl sulphoxide	30.20	29.8
Pyridine	47.16	33.1

With all the examples hydrogen bonding was either absent or neglegibly small. Both methods fail when strong hydrogen bonding or strong π-bonding is involved in the donor-acceptor interactions.

D. Hydrogen Bonding

Spectrophotometric measurements have been employed[28] to estimate the degree of interaction between vanadylbisacetylacetonate[28] which has one ligand site available, and various solvents. On coordination of the donor the optical spectrum is changed: band I is shifted towards longer wavelength and band II towards shorter wavelength. The empirical shift $D_{II, I}$ caused by a donor has been used to estimate the degree of interaction of the solvent D with vanadylbisacetylacetonate[28]. Taking the donor numbers of propanediol-1,2-carbonate and dimethylsulphoxide as reference points the $D_{II, I}$ values are found to be proportional to the respective donor numbers as long as hydrogen bridging is absent (Fig. 7).

The donor number calculated in this way gives reasonable results for some donor solvents. With the occurrance of hydrogen bonding the actual interaction is increased. Thus water is extremely strongly coordinated to the $V = O$ group not only by coordination at the V-atom, but also by hydrogen bridging at the oxygen of the $V = O$ group. This is confirmed by comparing the spectrophotometric data with calorimetric data, which show relatively small enthalpy changes for the interactions of alcohols with $[VO(acac)_2]$[29].

[28] SELBIN, J., and T. R. ORTOLANO: J. Inorg. Nucl. Chem. **26**, 37 (1964).
[29] CARLIN, R. L., and F. A. WALKER: J. Amer. Chem. Soc. **87**, 2128 (1965).

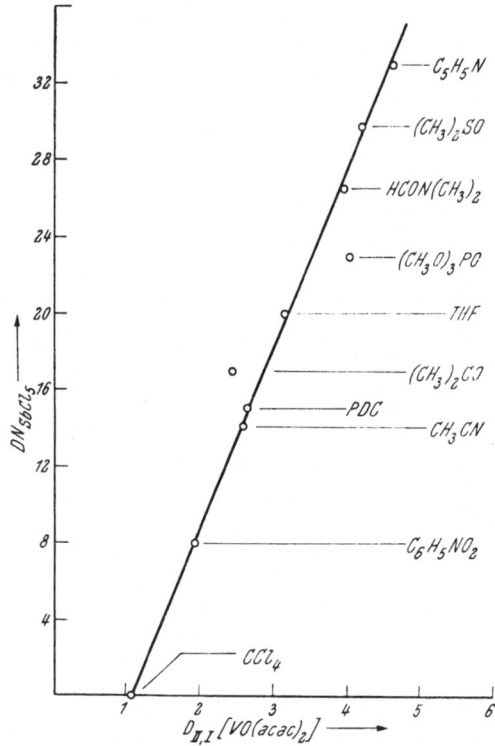

Fig. 7. Empirical shift $D_{II,I}$ [VO(acac)$_2$] and DN_{SbCl_5} of different solvents

Table 7. *Relationships between* $D_{II,I}$ *in Vanadylacetylacetonate by Coordination of One Solvent Ligand and the Donor Number*[30]

Solvent	$D_{II,I}$	approximate DN_{SbCl_5} (calc.)	DN_{SbCl_5} (found)
Benzene	1.58	4.9	
Nitrobenzene	1.95	8.4	2.7
Nitromethane	2.26	11.4	4.4
Acetonitrile	2.60	14.6	14.1
Propanediol-1,2-carbonate	2.65	15.1	15.1 (ref.)
Tetrahydrofurane	3.14	19.8	20.0
Dimethylformamide	3.94	27.5	26.6
Dimethyl sulphoxide	4.18	29.8	29.8 (ref.)
Pyridine	4.40	31.9	33.1
Trimethyl phosphate	4.02	28.2	23.0
Dioxane	4.02	28.2	
n-Butylalcohol	4.10	29.0	
Ethanol	4.26	30.4	
n-Propylamine	4.31	31.0	
Methanol	4.60	33.8	
Ammonia	4.62	34.0	
Formamide	5.15	39.1	
Water	5.49	42.3	18.0

[30] SELBIN, J.: Chem. Revs. **65**, 168 (1965).

It is also from this point of view, useful to distinguish hydrogen bonding solvents from those which cannot form hydrogen bridges to any appreciable extent. Although hydrogen bonding is not the same as hydrogen donating (protonic) it is generally true that protonic solvents are hydrogen bonded in the pure liquids to some extent.

E. Steric Factors

Donor molecules may be different in shape and size. When a single donor molecule is coordinated steric factors are likely to be of little importance, unless the available space at the coordination site is very small. Indeed steric contributions were not considered when the donor number was introduced. The situation is, however, different, when coordination of several solvent molecules occurs or when little room is available for the solvent molecules within the coordination sphere. Thus steric considerations will become important when small ions are to be coordinated by several large or bulky ligands. The results obtained will then not be in agreement with the donor number of the particular solvent.

The coordination of the vanadyl ion $[VO]^{2+}$ by five donor molecules to give $[VOD_5]^{2+}$ may serve as an example. Analysis of the spectra in acetonitrile, propanediol-1,2-carbonate, trimethylphosphate and dimethylsulphoxide reveals that coordination compounds with dimethylsulphoxide ($DN_{SbCl_5} = 29.8$) and acetonitrile ($DN_{SbCl_5} = 14.1$) have slightly distorted octahedral symmetries, whilst those with propanediol-1,2-carbonate ($DN_{SbCl_5} = 15.1$) and trimethylphosphate ($DN_{SbCl_5} = 23$) show a lower symmetry with higher π-bonding contributions to the V = O-bond. Thus interactions of the VO^{++} ion are stronger with dimethylsulphoxide and acetonitrile than with propanediol-1,2-carbonate and trimethylphosphate, which is not in agreement with their relative donor properties according to their donor numbers. Since the former require less room than the latter, these effects are likely to be mainly steric in nature.

With many other transition metal cations such steric considerations will be of importance to account for the observed behaviour.

F. Solubilities

Since dissolution processes of ionic or covalent compounds may be considered as donor-acceptor reactions, the donor number of the solvent will be of influence. Ionic compounds are likely to show increasing solubility with increasing donor number of the solvent. Steric factors must also be considered. Although acetonitrile and propanediol-1,2-carbonate have similar donor numbers, the former, which is small and rod-like in shape, shows higher solubilities for many compounds than the latter, which is larger and bulky. The higher dielectric constant of the latter does not appreciably contribute to solubilities, although it gives rise to solutions of higher conductivities than in acetonitrile.

In a qualitative way the solubility of nickel(II) chloride increases with increasing donor number of the solvent, e.g. Nitromethane < Acetonitrile < Water < Trimethyl phosphate < Dimethyl sulphoxide.

Thus a high donor number and favourable steric properties, are required to give useful solubilities; a high dielectric constant is not neccessary, but it is useful in order to keep the dissolved ions apart.

3. Coordination Equilibria in Solution

When an acceptor compound AX_n such as $SbCl_5$ is dissolved in a donor solvent D the formation of the solvate complex occurs according to [1]. The solvate complex may then be made to react with competitive ligands L. Such diplacement reactions can be carried out (a) by adding a stronger donor or (b) by disturbing the equilibrium, e.g. by differences in volatility, solubility or by addition of a large excess of the replacing donor. The formation of ammoniates of certain transition metal ions in aqueous solution occurs according to (a) due to the stronger donor properties of ammonia compared with water, although water with its lower volatility may be present in excess.

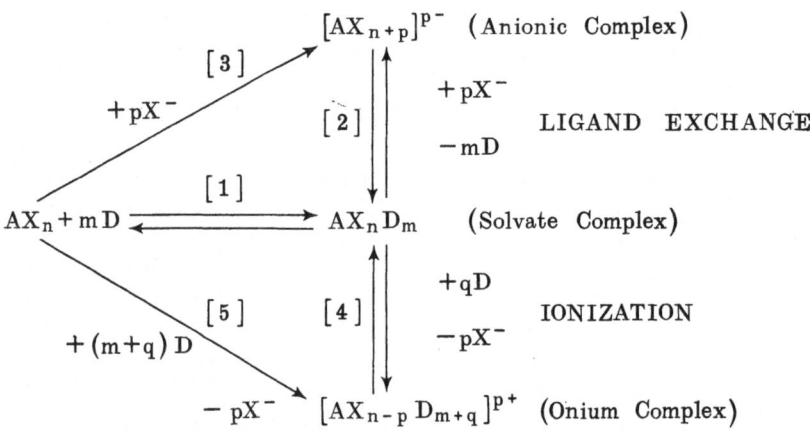

If L is an anion X^- the formation of an anionic complex will take place because solvent ligands D in the solvate complex will be replaced by X^- according to reaction [2]. The overall reaction is frequently simply represented by reaction [3] without regarding the intermediate steps described by [1] and [2].

The formation of the anionic complex is likely to occur if the competitive ligand X^- has stronger donor properties than the solvent molecules D. This condition will be best met by use of an acceptor solvent, where no competition is provided for coordination of L or X^- by the properties of the competitive ligand.

If the donor properties of the solvent are appreciably higher than those of the competitive ligand, substitution of X^- in the acceptor molecule or in the solvate by D will occur partly or even completely with the formation of solvated cations or of "Onium"-ions. The overall reaction is frequently simply represented by [5] and termed an ionization of the compound in the solvent. It is unfortunate that ionization and dissociation are frequently considered synonimous terms. In water such reactions are considered as hydrolysis:

$$MeX_n + 6\,H_2O \rightleftharpoons [Me(OH_2)_6]^{n+} + n\,X^-$$

While [1] + [2] lead to the formation of a complex ion, [1] + [4] lead to ionization, which is followed by dissociation according to the value of the dielectric constant of the solution.

A third possibility is that [1] is followed simultaneously by both [2] and [4] when reaction [4] may provide X^- units, which may be consumed by reaction [2]. This is likely to occur when the donor properties of the donor solvent and competitive ligand are not vastly different. This type of ionization is termed "auto-complex-formation" and is known to occur in many systems.

A. Formation of a (Non-solvated) Complex Anion

The following displacement reaction of the type [2] has been investigated for different donor solvents D:

$$(C_6H_5)_3CCl + D \cdot SbCl_5 \rightleftharpoons [(C_6H_5)_3C][SbCl_6] + D; \quad K_{SbCl_6^-}$$

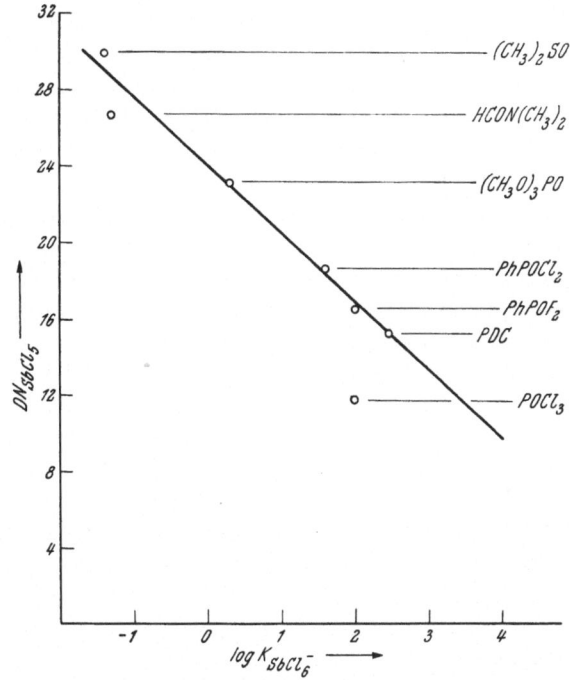

Fig. 8. $\log K_{SbCl_6^-}$-DN_{SbCl_5} plot in different solvents

The plot of the $\log K_{SbCl_6^-}$-values versus DN_{SbCl_5} of the respective donor molecules[1,26] D revealed the existence of the relationship[39]

$$\log K_{SbCl_6^-} = -a \cdot DN_{SbCl_5} + b$$

Thus the qualitative prediction has been confirmed that the formation constant of the anionic complex formed by the addition of the competitive anionic ligand X^- will increase with decrease in the donor number of the solvent used[1, 2, 26] (Table 8 and Fig. 8).

Table 8. *Formation Constants* $K_{SbCl_6^-}$ *of the Chloro-complexes from* $(C_6H_5)_3CCl$ *and* $SbCl_5$ *in Several Solvents of Different* DN_{SbCl_5} *at* $c \sim 10^{-3}$ *to* 10^{-5}

Solvent	DN_{SbCl_5}	$K_{SbCl_6^-}$
1,2-Dichloroethane	0.1	$> 10^5$
Sulphuryl chloride	0.1	$> 10^5$
Thionyl chloride	0.4	$> 10^5$
Benzoyl chloride	2.3	$> 10^5$
Phosphorus oxychloride	11.7	1.10^2
Acetonitrile	14.1	10^5
Propanediol-1,2-carbonate	15.1	3.10^2
Phenylphosphonic difluoride	16.4	1.10^2
Phenylphosphonic dichloride	18.5	4.10^1
Trimethyl phosphate	23.0	2.10^0
Dimethylformamide	26.6	5.10^{-2}
Dimethyl sulphoxide	29.8	4.10^{-2}
Pyridine	33.1	$< 10^{-2}$

With a given formation constant of the anionic complex K_3, strong donor properties of the solvent D will lead to low values of $K_{SbCl_6^-}$ since the formation constant of the solvate complex K_1 will be high. On the other hand high values for $K_{SbCl_6^-}$ are expected in a solvent of low DN_{SbCl_5}. It is apparent that efforts must be made in order to estimate, in at least a semiquantitative way, the donor properties of a number of competitive ligands.

Thus a solvent of high donor number will not be useful for the formation of bromo- and chloro-complexes of group (a) metals, although it may serve its purpose as a reaction medium for the formation of stronger complexes such as cyano-complexes. It is now clear why so many complex compounds are readily prepared in water, which has a medium donor number of approximately $DN_{SbCl_5} = 18$ and why many reactions depend on the p_H-value (high donor properties of the hydroxyl-ions). Fluoro-, azido-, cyano- and thiocyanato-complexes of group (a) metal ions are readily formed, since the donor properties of water are lower than those of the competitive ligands. On the other hand water is a poor medium for the formation of chloro-, bromo- and iodo-complexes of class (a) metal ions. Some of them are only formed in the presence of a large excess of the competitive ligand, whilst others are not formed at all in this medium.

Non-coordinating solvents such as carbon tetrachloride would be useful media for the formation of various complexes, if they would provide sufficient solubilities for the reactants.

Certain acceptor solvents are very useful for the formation of anionic complexes provided that the reactants can be brought into solution. Thus liquid sulphur dioxide has been used for the formation of certain chloro-complexes such as $K^+[SbCl_6]^-$, $[CH_3CO]^+[SbCl_6]^-$, $[C_6H_5CO]^+[SbCl_6]^-$, $[C_6H_5CO]^+[AlCl_4]^-$, $[NO]^+[SbCl_6]^-$, $[NO_2]^+[SbCl_6]^-$ and others[31-33]. Liquid hydrogen fluoride[34] or

[31] SEEL, F., and H. BAUER: Z. Naturf. **2 b**, 397 (1947).

[32] SEEL, F., J. NOGRADI, and R. ROSSE: Z. anorg. allg. Chem. **269**, 197 (1952).

[33] SEEL, F.: Z. anorg. allg. Chem. **250**, 331 (1943); **252**, 24 (1943).

[34] HYMAN, H. H., and J. J. KATZ: Chapter II in "Non Aqueous Solvent Systems," Ed. T. C. WADDINGTON, Academic Press, London, New York 1965.

bromine(III) fluoride[35] are convenient media for the preparation of a large number of fluoro-complexes since anionic ligands are supplied by the self-ionization equilibria, which act as powerful sources of fluoride ions:

$$3\ HF \rightleftharpoons [H_2F]^+ + [HF_2]^-; \quad [HF_2]^- \rightleftharpoons HF + F^-$$
$$2\ BrF_3 \rightleftharpoons [BrF_2]^+ + [BrF_4]^-; \quad [BrF_4]^- \rightleftharpoons BrF_3 + F^-$$

Certain acid chlorides, such as nitrosyl chloride serve a similar purpose in the formation of chloro-complexes due to the availability of chloride ions in the pure solvent[36].

$$NOCl \rightleftharpoons NO^+ + Cl^-$$

Thus nitrosonium chlorometallates are formed even in the absence of a solute which provides chloride ions[36, 37]:

$$FeCl_3 + NOCl \rightleftharpoons NO^+ + [FeCl_4]^-$$

Likewise, chlorometallates are known to be formed in iodine monochloride solutions[38].

Thionyl chloride[39, 40] and sulphuryl chloride[41, 42] have exceedingly low donor numbers[26] with good solubility properties for covalent acceptor chlorides as well as for tetra-alkylammonium chloride, and chloro-complexes are readily formed in their solutions:

$$2\ (C_2H_5)_4NCl + SnCl_4 \rightarrow [(C_2H_5)_4N]_2[SnCl_6]$$

A stoichiometric amount of R_4NCl will usually lead to the quantitative conversion of the acceptor chloride into the respective chloro-complex. On the other hand owing to their small DN_{SbCl_5} they behave as poor solvents for most ionic compounds and also for most transition metal halides. Alkali or alkaline earth salts of the chloro-complexes cannot be formed from their solutions and chloro-complexes of most transition metal ions are also inaccessible in such media. Their use is therefore restricted to the formation of chloro-complexes of certain representative elements with large cations, such as $[R_4N]^+$.

Similar solvent properties are exhibited by acetyl- or benzoyl chloride[43-45] as solvents, while phosphorus oxychloride[46], phenylphosphonic dichloride[46] and selenium oxychloride[47] have somewhat higher donor numbers. They are still useful for the preparation of various complex chlorides, in particular selenium oxychloride, which has also a reasonable dielectric constant. Ionic compounds such as potassium chloride can be dissolved with the rapid formation of chloro-complexes:

$$KCl + SeOCl_2 \rightleftharpoons K^+SeOCl_3^-$$

[35] SHARPE, A. G.: Chapter VII in "Non-Aqueous Solvent Systems," Ed. T. C. WADDINGTON, Academic Press, London, New York 1965.
[36] BURG, A. B., and G. W. CAMPBELL: J. Amer. Chem. Soc. **70**, 1964 (1948).
[37] BURG, A. B., and D. E. McKENZIE: J. Amer. Chem. Soc. **74**, 3243 (1954).
[38] GUTMANN, V.: Z. anorg. allg. Chem. **264**, 156 (1951).
[39] SPANDAU, H., and E. BRUNNECK: Z. anorg. allg. Chem. **270**, 201 (1952).
[40] SPANDAU, H., and E. BRUNNECK: Z. anorg. allg. Chem. **278**, 197 (1955).
[41] GUTMANN, V.: Mh. Chem. **85**, 393 (1954).
[42] GUTMANN, V.: Mh. Chem. **85**, 404 (1954).
[43] GUTMANN, V., and H. TANNENBERGER: Mh. Chem. **88**, 216, 292 (1957).
[44] GUTMANN, V., and G. HAMPEL: Mh. Chem. **92**, 1048 (1961).
[45] SINGH, J., R. C. PAUL, and S. S. SANDHU: J. Chem. Soc. **1959**, 846
[46] GUTMANN, V.: Oesterr. Chem. Ztg. **62**, 326 (1961).
[47] SMITH, G. B. L.: Chem. Revs. **23**, 165 (1938).

Anionic complexes other than chloro-complexes cannot be expected in such solvents due to the occurrance of solvolysis.

When the formation of bromo-complexes is required, iodine monobromide[48], mercuric bromide[49, 50] or an oxybromide, such as benzoyl bromide[51, 52], may be used as solvents. It is possible that nitrosyl bromide may serve the same purpose.

This group of solvents is, however, limited in its applications. Other solvents, such as chloroform or benzene have been used for certain reactions involving covalent compounds. Thus triphenylchloromethane has been used as a source of chloride ions to combine with certain metal chlorides in solutions of acetic acid[53], nitroalkanes[54, 55], chlorobenzenes[55] and benzene[55]. All of these reactions cannot be carried out in solvents of high donor numbers such as dimethyl sulphoxide or tributylphosphate[56].

The most versatile solvents for the purpose under consideration will have medium donor numbers, no self-ionization producing a certain type of anionic ligands and a reasonable dielectric constant. In particular acetonitrile ($DN_{SbCl_5} = 14$, $DEC = 37$), propanediol-1,2-carbonate ($DN_{SbCl_5} = 15$, $DEC = 69$) or sulpholane ($DN_{SbCl_5} = 14.3$, $DEC = 42$) are to be recommended. Numerous bromo-, chloro-, azido-, cyano-, thiocyanato- and other complexes of class (a) metals are readily formed in their solutions[1].

Certain differences in the properties of acetonitrile and propanediol-1,2-carbonate (both have similar donor numbers) can be attributed to steric influences due to the different sizes and shapes of the solvent molecules. Acetonitrile, being smaller in size and more favourable in shape as a ligand, is a better solvent for many ionic compounds and assists the formation of anionic complexes sometimes more readily than propanediol-1,2-carbonate, although the latter has a higher dielectric constant.

Complex compounds with sufficiently strong ligands can also be obtained in solvents of high donor number, such as trimethylphosphate. It would be extremely useful to obtain quantitative data on the donor properties of competitive ligands. (see also p. 168).

B. Ionization and Dissociation of a Solvate Complex

If the solvent is an appreciably stronger donor than the ligand of the dissolved compound, replacement by solvent molecules of the ligand X^- in the acceptor molecule AX_n will take place. If X^- is negatively charged positive ions will be formed in the solutions. Again it is evident that a high dielectric constant will favour dissociation following ionization. The main requirements for the solvent

[48] GUTMANN, V.: Mh. Chem. **82**, 156 (1951).

[49] JANDER, G., and K. BRODERSEN: Z. anorg. allg. Chem. **261**, 261 (1950).

[50] JANDER, G., and K. BRODERSEN: Z. anorg. allg. Chem. **262**, 22 (1950).

[51] GUTMANN, V., and K. UTVARY: Mh. Chem. **89**, 186, 732 (1958).

[52] GUTMANN, V., and K. UTVARY: Mh. Chem. **90**, 710 (1959).

[53] COTTON, J. L., and A. G. EVANS: J. Chem. Soc. **1959**, 2988.

[54] BAYLES, J. W., A. G. EVANS, and J. R. JONES: J. Chem. Soc. **1955**, 206.

[55] BAYLES, J. W., A. G. EVANS, and J. R. JONES: J. Chem. Soc. **1957**, 2020.

[56] BAAZ, M., V. GUTMANN, and J. R. MASAGUER: Mh. Chem. **92**, 590 (1961).

to achieve reaction [4] are high donor number (stronger donor properties than those of the X⁻-ligand), reasonable dielectric constant and small size of the solvent molecule. The replacement of the ligand X⁻ in the solvate complex by solvent molecules can be assisted by the presence of a strong acceptor for the ligand. Under such conditions solvents of only moderate donor number can be used such as phosphorus oxychloride in the presence of antimony(V) chloride[57]:

$$Cl_4Ti(OPCl_3)_2 + Cl_5SbOPCl_3 \rightleftharpoons [Cl_3Ti(OPCl_3)_3]^+ + [SbCl_6]^-$$

Table 9. *Onium Ions in Certain Non-Aqueous Solutions*

Ion \ Solvent	POCl₃	AN	PDC	C₆H₅POCl₂	TMP	DMSO
[MnBr]⁺		+				
[MnCl]⁺		+				
[CoBr]⁺		+			+	
[CoN₃]⁺					+	
[NiBr]⁺		+				
[CuCl]⁺		+			+	+
[CuN₃]⁺		+			+	+
[VOCl]⁺		+	+			+
[VON₃]⁺		+	+			+
[ZnCl]⁺	+	+		+		+
[HgCl]⁺	+			+		
[TiCl]²⁺			+			
[TiCl₂]⁺					+	
[Ti(N₃)₂]⁺					+	
[VBr]²⁺					+	
[VBr₂]⁺		+				
[VCl]²⁺			+		+	
[VN₃]²⁺		+	+		+	
[CrCl]²⁺					+	
[CrCl₂]⁺		+	+			
[CrN₃]²⁺					+	
[Cr(N₃)₂]⁺		+	+			
[FeCl]²⁺		+	+			
[FeCl₂]⁺					+	
[FeN₃]²⁺			+			
[Fe(N₃)₂]⁺			+			
[FeF₂]⁺						+
[Fe(CN)₂]⁺						+
[BCl₂]⁺	+	+		+		
[AlCl]²⁺	+			+		
[SeCl₃]⁺	+	+		+		
[SnCl₃]⁺	+			+		
[TiCl₃]⁺	+			+		
[ZrCl₃]⁺	+					
[PCl₄]⁺	+	+		+		
[SbCl₄]⁺		+				
[NbCl₄]⁺	+					
[TaCl₄]⁺	+					

[57] AGERMANN, M., L. H. ANDERSSON, I. LINDQVIST, and M. ZACKRISSON: Acta Chem. Scand. **12**, 477 (1958).

Since water meets these requirements for many compounds, it is an excellent ionizing solvent. Frequently the replacement of ligands by water molecules takes place so readily that no intermediate steps such as halonium compounds are formed.

When aluminium trichloride is dissolved in water, the complete replacement of chloride by water molecules with additional coordination of water molecules takes place and polynuclear hydrated Al^{3+} is finally found in its solution. Intermediate species, such as $[AlCl]^{2+}_{solv.}$, can be detected in phosphorus oxychloride or in phenyl phosphonic dichloride[58] in the presence of antimony(V) chloride or ferric chloride[46]. Other chloronium ions are formed in solvents of moderate donor number in the presence of a suitable acceptor[8].

$$NCl_n + p\,MCl_m \rightleftharpoons [NCl_{n-p}]^{p+} + p\,[MCl_{m+1}]^-$$

In a solvent of high DN_{SbCl_5} such ionization is complete and all metal-halide bonds are replaced by metal-solvent bonds. Dimethyl sulphoxide (DMSO, $DN_{SbCl_5} = 30$) is a solvent favouring ionization of halides. Ferric chloride does not accept chloride ions to give $[FeCl_4]^-$ in this solvent even in the presence of a large excess chloride ions. The conducting solutions contain $[Fe(DMSO)_6]^{3+}$ and chloride ions[59, 60]. The same is true for its solutions in tributylphosphate.

$$Cl_3Fe(DMSO)_3 + 3\ DMSO \rightleftharpoons [Fe(DMSO)_6]^{3+} + 3\ Cl^-$$

The crystalline dimethylsulphoxide solvate of vanadyl chloride $[VOCl_2(DMSO)_3]$ gives in DMSO the spectrum of $[VOCl]^+_{solv.}$ and addition of chloride ions will give neither the spectrum of the undissociated dichloride nor spectra of anionic chloro-complexes. Thus one chloride is readily ionized in DMSO even in the presence of excess chloride ions[61].

$$[(DMSO)_3VOCl_2] + DMSO \rightleftharpoons [(DMSO)_4VOCl]^+ + Cl^-$$

It is expected that vanadyl bromide will completely ionize in dimethylsulph-oxide or tributylphosphate, but the situation may be different for fluorides be-cause the fluoride ion has stronger donor properties than bromide or chloride ions towards class (a) metals. Ionization of cyanides and thiocyanates of class (a) metals is unlikely in solvents of donor number below 25; iodides and bromides of class (a) metal ions are completely ionized in solvents of high donor numbers, whilst iodides and bromides of class (b) metals are not ionized in the same media.

C. Autocomplex Formation

When the coordinating properties of the donor solvent and the ligand are not vastly different reactions [2] and [4] may both follow reaction [1]. Thus the solvate complex will partly gain solvent molecules with simultaneous release of ligand

[58] BAAZ, M., V. GUTMANN, L. HÜBNER, F. MAIRINGER, and T. S. WEST: Z. anorg. allg. Chem. **311**, 302 (1961).

[59] GUTMANN, V., and L. HÜBNER: Mh. Chem. **92**, 1261 (1961).

[60] GUTMANN, V., and G. HAMPEL: Mh. Chem. **94**, 830 (1963).

[61] GUTMANN, V., and H. LAUSSEGGER: Mh. Chem. **98**, 439 (1967).

ions which can be used to replace solvent molecules in other solvate complexes to give complex anions. This process is called autocomplex formation.

While the formation of an anionic complex is supported by a low donor number of the solvent and by high donor properties of X^-, the formation of solvated cations (ionization) is favoured by a high donor number of the solvent. Autocomplex formation will be expected to occur, when the donor properties of solvent and anionic ligand are similar.

Examples of autocomplex formation can be seen from Table 10, where A^- (= anionic complex) denotes the ready formation of complex anions in the presence of the competitive anionic ligand X^-.

Autocomplex formation with a given cation occurs generally in a solvent of high donor number when the donor properties of the competitive ligand are also high. The bromide ion is a weaker ligand towards Co^{++} than the chloride ion. Autocomplex formation of $CoBr_2$ is considerable in propanediol-1,2-carbonate and trimethylphosphate[62], which have lower donor numbers than dimethylsulphoxide in which $CoCl_2$ is found to undergo autocomplex formation[59].

Table 10. *Examples of the Qualitative Behaviour of Certain Systems in Solvents of Different Donor Number (Donor Numbers in Parenthesis)* A^-: *Formation of Anionic Complexes*

System \ Solvent (DN)	POCl₃ (11)	AN (14)	PDC (15)	TMP (23)	DMSO (30)
$Sb^{5+} + Cl^-$	A^-	Autocomplex		Ionization	Ionization
$Co^{2+} + Br^-$		A^-	Autocomplex	Autocomplex	Ionization
$Co^{2+} + Cl^-$		A^-		A^-	Autocomplex
$Fe^{3+} + Cl^-$	A^-	A^-	A^-	Autocomplex	Ionization
$Fe^{3+} + N_3^-$		A^-		A^-	Autocomplex
$Fe^{3+} + CN^-$		A^-		A^-	A^-
$VO^{2+} + Cl^-$		A^-	A^-	Autocomplex	Ionization
$VO^{2+} + N_3^-$		A^-	A^-	A^-	Autocomplex

A more detailed discussion has been attempted in Chapter VIII.

Similar results are found for iron(III) compounds. While $FeCl_3$ undergoes autocomplex formation in triethylphosphate[4] ($DN_{SbCl_5} = 23$) it is simply ionized in a solvent of higher DN_{SbCl_5}, such as dimethyl sulphoxide[59, 60]. Ferric chloride prefers to form tetrachloroferrate in a solvent of much lower donor number, for example phosphorus oxychloride[46] ($DN_{SbCl_5} = 11$). With a stronger competitive ligand, such as azide, autocomplex formation is found in dimethyl sulphoxide, where with the stronger donating cyanide ion anionic cyanocomplexes are easily formed[63].

4. Solvation and Donor Properties

Solvation of the dissolved species is an extremely important phenomenon in solution. It involves both coordination of solvent molecules and additional attachment of the latter to the dissolved particles, with distinctly weaker interactions.

---

[62] GUTMANN, V., and K. FENKART: Mh. Chem. **98**, 1 (1967).
[63] CSISZAR, B., V. GUTMANN, and E. WYCHERA: Mh. Chem. **98**, 12 (1967).

Since the donor number of the solvent characterizes the magnitude of the co-ordination of the donor solvent by an electron pair acceptor, it was of interest to investigate the relations between donor number and solvation[64].

The relation between the standard free energy ΔG_0 of the reaction $M(s) \rightarrow M^{z+}(solv.)$ and the standard electrode potential E_0 for the system $M(s)/M^{z+}(solv.)$ is given by the equation $\Delta G_0 = -z \cdot F \cdot E_0$. For the estimation of the standard electrode potential of a metal ion, a Born-Haber cycle consisting of the following 3 steps may be considered: *a)* sublimation of the metal, *b)* ionization of the gaseous metal atom and *c)* solvation of the ion.

For a given metal ion the energies for *(a)* and *(b)* remain constant and hence the value for E_0 in various solvents is determined by the corresponding free energies of solvation.

For metals which are soluble in mercury (e.g. the alkali and alkaline earth metals) the polarographic half-wave potential $E_{1/2}$ is a function of (i) the standard electrode potential of the metal-metal ion couple, (ii) the solubility of the metal in mercury and (iii) the free energy of amalgamation. (ii) and (iii) are independent of the nature of the solvent.

Thus, for a reversible reduction the half-wave potential is a measure of the interaction of the metal ion with solvent molecules according to the reaction:

$$M^{z+}(sv) + ze^- \rightarrow M(Hg) + solvent$$

For an irreversible reduction the half-wave potential is determined not only by the standard electrode potential, but also by the polarographic overvoltage. For a simple electrode process the metal ion-solvent interaction is mainly respon-sible for the polarographic overvoltage, and hence $E_{1/2}$ of simple irreversible reductions may also be considered as a function of the solvation.

In order to compare the half-wave potentials of a given metal ion in different solvents (*a*) measurements must be made versus a defined reference electrode, such as the aqueous saturated calomel electrode and (*b*) the liquid junction poten-tial must be eliminated. The latter can be achieved with reasonable approximation by applying the method of the reference ion, as suggested by PLESKOV[65]; the assumption is made that the solvation energy of Rb^+ is practically constant in all solvents, i.e. the $E_{1/2}$ of Rb^+ is nearly constant in all solvents. Complexation is minimized by using perchlorates as supporting electrolytes, and the effect of ion pair formation can be neglected, when solvents of medium or high dielectric constant ($\varepsilon > 20$) are considered.

No general relation between the shift of the half-wave potential in different solvents and their physical properties such as dielectric constant or viscosity has been found[66-70], but the solvating properties of a solvent play a prominent role.

[64] GUTMANN, V., G. PEYCHAL-HEILING, and M. MICHLMAYR: Inorg. Nucl. Chem. Letters, **3**, 501 (1967).
[65] PLESKOV, V. A.: Uspekhi Khim, **16**, 254 (1947).
[66] LARSON, R. C., and R. T. IWAMOTO: J. Am. Chem. Soc. **82**, 3239, 3526 (1960).
[67] TAKAHASHI, R.: Talanta **12**, 1211 (1965).
[68] KOLTHOFF, I. M.: J. Pol. Soc. **10**, 22 (1964).
[69] COETZEE, J. F., and D. K. McGUIRE: J. Phys. Chem. **67**, 1814 (1963).
[70] BRUSS, D. B., and T. DE VRIES: J. Am. Chem. Soc. **78**, 733 (1956).

The comparison of the half-wave potentials of alkali and alkaline earth metal ions expressed in the "rubidium scale" in various solvents (Table 11) shows that a relation exists between $E_{1/2}$ and the donor number of the solvent, rather than the dielectric constant (see Table 12). Although the dielectric constants of CH_3CN and PDC are different, the half-wave potential of a given ion is similar in both solvents as well as their donor numbers. CH_3CN and DMA have nearly identical dielectric constants, but the latter has a much higher donor number. Indeed $E_{1/2}$ is more negative in DMA than in CH_3CN indicating a higher stability of the solvate complex in DMA.

Table 11. *Half-wave Potentials of Alkali and Alkaline Earth Metal Ions in Different Solvents in Volts versus Aqueous Saturated Calomel Electrode and versus the Half-wave Potential of* Rb^+ *(below and in italics)*

Solvent Ref.	DMSO [71]	DMA [72]	DMF¹ [73]	Water [74]	Acetone [75]	PDC [76]	CH₃CN [77]	Benzonitrile [66]
Li^+	−2.45	−2.38	−1.81	−2.33		−1.99	−1.95	−1.82
	−0.39	*−0.34*	*−0.29*	*−0.20*		*−0.02*	*+0.03*	*+0.06*
Na^+	−2.07	−2.06	−1.53	−2.12	−1.92	−1.96	−1.85	−1.74
	−0.01	*−0.02*	*−0.01*	*+0.01*	*+0.05*	*+0.01*	*+0.13*	*+0.14*
K^+	−2.11	−2.08	−1.55	−2.14	−1.97	−1.84	−1.96	
	−0.05	*−0.04*	*−0.03*	*−0.01*	*0.00*	*+0.13*	*+0.02*	
Rb^+	−2.06	−2.04	−1.52	−2.13	−1.97	−1.97	−1.98	−1.88
	0.00	0.00	0.00	0.00	0.00	0.00	0.00	0.00
Cs^+	−2.03	−2.03	−1.53	−2.09		−1.97	−1.97	
	+0.03	*+0.01*	*−0.01*	*+0.04*		0.00	*+0.01*	
Mg^{2+}	−2.28	−2.30		−2.20	−1.75	−1.72	−1.84	−1.62
	−0.22	*−0.26*		*−0.07*	*+0.22*	*+0.25*	*+0.14*	*+0.26*
Ca^{2+}	−2.30	−2.37	−1.84	−2.20		−1.92	−1.82	−1.73
	−0.24	*−0.33*	*−0.31*	*−0.07*		*+0.05*	*+0.16*	*+0.15*
Sr^{2+}	−2.10	−2.23	−1.64	−2.11		−1.83	−1.76	−1.72
	−0.04	*−0.19*	*−0.12*	*+0.02*		*+0.14*	*+0.22*	*+0.16*
Ba^{2+}	−2.09	−2.02	−1.49	−1.92		−1.67	−1.63	−1.58
	−0.03	*+0.02*	*+0.03*	*+0.21*		*+0.30*	*−0.35*	*+0.30*

¹ Versus mercury pool as reference electrode.

The half-wave potential of Cs^+ is almost independent of the nature of the solvent as well as that postulated for Rb^+. Na^+ and K^+ still have low solvation energies, but they vary with the nature of the solvent, showing a decrease with decreasing donor number. A stronger dependence is found for Li^+, which is known to have stronger solvating tendencies than the other alkali metal ions.

71 SCHÖBER, G., and V. GUTMANN: Adv. in Polarography, Ed. I. S. LONGMUIR, p. 940, Pergamon Press, London 1960.
72 GUTMANN, V., G. PEYCHAL-HEILING, and M. MICHLMAYR: Anal. Chem. in press.
73 BROWN, G. H., and R. AL-URFALI: J. Am. Chem. Soc. **80**, 2113 (1958).
74 MEITES, L.: "Polarographic Techniques," Interscience, New York 1955.
75 COETZEE, J. F., and SIAO WEI-SAN: Inorg. Chem. **2**, 14 (1963).
76 GUTMANN, V., M. KOGELNIG, and M. MICHLMAYR: Mh. Chem. in press.
77 KOLTHOFF, I. M., and J. F. COETZEE: J. Am. Chem. Soc. **79**, 870 (1957).

Table 12. *Dielectric Constants and Donor Numbers of Various Solvents*

Solvent	Dielectric Constant	Donor Number
DMSO	45	29.8
DMA	37.8	27.8
DMF	36.7	26.6
Water	81	18
Acetone	20.7	17
PDC	69	15.1
CH_3CN	38.8	14.1
Benzonitrile	25.2	11.9

With stronger solvation of an ion (alkaline earth metal ions) the change of the half-wave potentials with change of solvent becomes even more pronounced.

This relationship has also been found by comparing the half-wave potentials of vanadium(III)-, vanadyl- and chromium(III) ions in water, dimethylformamide and dimethyl sulphoxide[78].

Quantitative conclusions cannot be drawn as long as no data are available which have been obtained under strictly analogous experimental conditions, and as long as other contributions to the data, such as the activity coefficients of the electroactive species and of the supporting electrolyte, and microscopic effects (the structure of the solvent and the dielectric constant in the vicinity of the ion) are not taken into account. The relationship between the half-wave potentials of a given ion in different solvents and the donor number of the solvent is, however, striking and suggests that, indeed, with increasing donor properties of the solvent its solvating properties become stronger.

It has been shown in recent years that definitive solvation numbers may be obtained for certain metal cations in water[79], methanol[80], liquid ammonia[81], N, N-dimethylformamide[82] and dimethyl sulphoxide[83] by use of NMR-techniques. For the Mg^{++} in liquid ammonia the unexpected solvation number of 5 has been found[81]. In methanol six solvent molecules were found associated, with each magnesium ion in the primary solvation sphere. Since only one LORENTZIAN signal was observed for bound ammonia, the exchange between non equivalent ammonias must be rapid. The most important factor in the formation of the pentaammoniated magnesium ion has been assumed to be rather strong ion pairing, which is known to occur in liquid ammonia[81].

[78] MICHLMAYR, M., and V. GUTMANN: Inorg. Chim. Acta, **1**, 471 (1967).
[79] CONNICK, R. E., and D. N. FIAT: J. Chem. Phys. **39**, 1349 (1963).
[80] SWINEHART, J. H., and H. TAUBE: J. Chem. Phys. **37**, 1579 (1962).
[81] SWIFT, T. J., and H. H. LO: J. Am. Chem. Soc. **89**, 3988 (1967).
[82] MATWIYOFF, N. A.: Inorg. Chem. **5**, 788 (1966).
[83] THOMAS, S., and W. L. REYNOLDS: J. Chem. Phys. **44**, 3148 (1966).

Chapter III

Coordination Chemistry in Proton-containing Donor Solvents

1. General Properties of Proton-containing Solvents

The most important structural feature of proton-containing solvents is the occurrance of hydrogen bonding in the liquid states. The strength of the hydrogen bond varies from solvent to solvent and the degree of association is also characteristic for each solvent. Since hydrogen bonding in these solvents is essentially due to weak electrostatic interactions between the solvent dipoles, the strongest interaction occurs in liquid hydrogen fluoride (8 kcal/Mole) where zig-zag chains are (probably) present due to strong hydrogen bridging between the solvent molecules

Hydrogen bonding is weaker in liquid water (5 kcal/Mole) where oxygen is tetracoordinated and still weaker in liquid ammonia. The hydrides of the elements of the following period, namely PH_3, H_2S and HCl show extremely weak tendencies to hydrogen bonding, as may be seen from a comparison of freezing points and boiling points.

Association in HF, H_2O and NH_3 is also maintained when "solvent-ions" are produced such as may occur by the self-ionization equilibria, usually represented according to the equations:

$$2\ HF \rightleftharpoons H_2F^+ + F^-$$
$$2\ H_2O \rightleftharpoons H_3O^+ + OH^-$$
$$2\ NH_3 \rightleftharpoons NH_4^+ + NH_2^-$$

It has been customary to indicate only the solvation of the proton in terms of the smallest units possible, but H_2F^+ may better be represented as $[(HF)_nH]^+$ and F^- as $[(HF)_mF]^-$.

The ease of occurrance of proton transfer reactions is also demonstrated by the apparently high mobilities of the solvent ions. Due to a chain-like mechanism, the essentials of which were first described by GROTTHUS in 1806, the proton is attached to an associated solvent unit, which in turn may transfer another proton to another bulk of solvent molecules. The high mobilities of the solvent anions are simulated by proton transfer reactions: a hydrogen bridged solvent unit may loose a proton to one of its neighbours leaving a negatively charged unit. This may accept a proton, thus leaving the proton donating species with a negative charge.

Liquid ammonia appears to be an exception to this rule. The conductivities of solutions of acids and bases are high, but so are the conductivities of other electrolyte solutions in this solvent. An explaination has been advanced from a structural point of view: all units involved in frequent proton transfer reactions are not fully coordinated and polar such as H_3O^+, H_2O and OH^- in water or H_2F^+, HF and F^- in liquid hydrogen fluoride. In contrast the NH_4^+ ion produced in liquid ammonia is symmetrical and non-polar and release of a proton is not easily possible[1] (the NH_4^+-ion is a very weak Brönsted acid, while the H_3O^+-ion is the strongest Brönsted acid in water).

High mobilities of the solvent ions are found in anhydrous sulphuric acid, hydrofluoric acid and to a considerably smaller extent for the solvent cations in alcohols, such as methyl and ethyl alcohol.

$$2 H_2SO_4 \rightleftharpoons H_3SO_4^+ + HSO_4^-$$
$$2 CH_3OH \rightleftharpoons CH_3OH_2^+ + CH_3O^-$$
$$2 C_2H_5OH \rightleftharpoons C_2H_5OH_2^+ + C_2H_5O^-$$

Other protonic solvents are nitric acid, carboxylic acids such as formic and acetic acids, as well as hydrazine, hydrogen cyanide and to a limited extent hydrogen sulphide and the other hydrogen halides:

$$2 HNO_3 \rightleftharpoons H_2NO_3^+ + NO_3^-$$
$$2 HCOOH \rightleftharpoons HCOOH_2^+ + HCOO^-$$
$$2 CH_3COOH \rightleftharpoons CH_3COOH_2^+ + CH_3COO^-$$
$$2 N_2H_4 \rightleftharpoons N_2H_5^+ + N_2H_3^-$$
$$2 HCN \rightleftharpoons H_2CN^+ + CN^-$$
$$2 H_2S \rightleftharpoons H_3S^+ + SH^-$$
$$3 HCl \rightleftharpoons H_2Cl^+ + HCl_2^-$$

In all proton containing solvents acid-base phenomena can be described in terms of the BRÖNSTED-LOWRY theory. All of these solvents have the solvated proton in common as the solvent cation, and this determines to a considerable extent the chemistry in their solutions. Brönsted acids are usually characterized by their acidic strength in water, e.g. by the acidity constant in this solvent. Thus acetic acid and hydrofluoric acid both behave as moderately weak acids in water with $K \sim 10^{-5}$ at room temperature. When acetic acid is dissolved in liquid hydrogen fluoride, the former is successfully competing for the protons, so that acetic acid acts as a base ("acetic-base") in this medium just as it does in nitric acid:

$$CH_3COOH + HF \quad \rightleftharpoons CH_3COOH_2^+ + F^-$$
$$CH_3COOH + HNO_3 \rightleftharpoons CH_3COOH_2^+ + NO_3^-$$
base I   acid II   acid I   base II

Although nitric acid is in water a much stronger acid than hydrofluoric acid, it acts as a base when dissolved in liquid hydrogen fluoride:

---
[1] GURNEY, R. W.: "Ionic Processes in Solution," McGraw Hill, London 1953.

$$HNO_3 \; + \; HF \; \rightleftharpoons \; H_2NO_3^+ \; + \; F^-$$
base I   acid II   acid I   base II

$HClO_4$ appears to be acidic in most other "acidic systems":

$$HClO_4 + CH_3COOH \rightleftharpoons ClO_4^- + CH_3COOH_2^+$$
$$HClO_4 + HF \qquad \rightleftharpoons ClO_4^- + H_2F^+$$
$$HClO_4 + HNO_3 \qquad \rightleftharpoons ClO_4^- + H_2NO_3^+$$
acid II   base I   base II   acid I

Another example of the marked influence of the relative proton affinities is the behaviour of acetamide, which is known to behave as a base in water and as an acid in liquid ammonia:

in water:
$$CH_3CONH_2 + H_2O \rightleftharpoons CH_3CONH_3^+ + OH^-$$
      base   acid

in ammonia:
$$CH_3CONH_2 + NH_3 \rightleftharpoons CH_3CONH^- + NH_4^+$$
    acid   base

In general it is true that a solvent with a high proton affinity, such as liquid ammonia or hydrazine, will be a levelling solvent for acids, while a solvent of low proton affinity such as acetic acid, hydrogen fluoride, and nitric or sulphuric acid will be a levelling solvent for bases.

Certain organic nitrogen bases show exceedingly small basic properties in water, so that they cannot be determined by acidimetric titrations; they can, however, readily be titrated in solutions of acetic acid using perchloric acid, since acidic solvents are levelling solvents for bases. It is obvious that in most protonic solvents normal p_H-color indicators can be used to follow acid-base reactions.

Frequently the statement is found that acetic acid is a stronger acid in liquid ammonia than it is in water. It is, however, not true that the dissociation constant of acetic acid is higher in liquid ammonia than it is in water. Comparison of dissociation constants, which are indicative of the acidity of a compound in a particular solvent, is only meaningful in the same solvent. Comparison of dissociation constants in different solvents just have no meaning. In water H_3O^+-ions are produced by the addition of a quantity of acetic acid just as NH_4^+-ions are produced in liquid ammonia in the presence of acetic acid, but the differences in dielectric constants make the actual concentration of NH_4^+-ions in liquid ammonia lower than that of H_3O^+-ions in water at the same concentration. The H_3O^+-ion shows stronger acidic properties than the NH_4^+-ion.

Table 13. *Autoprotolysis Constants* K *for Different Solvents at 25°*

Solvent	$-\log K$
Ammonia	~ 30
Ethyl alcohol	19
Acetic acid	~ 13
Hydrogen fluoride	~ 10
Formic acid	6
Sulphuric acid	3.6
Phosphoric acid	2

The differences in autoprotolysis constants (Table 13) suggest that different p_H-ranges are available for acid-base reactions in the respective solvents.

It must be born in mind that owing to the highly acidic properties of the H_3O^+-ion—the strongest acid known in aqueous solutions—the p_H is approximately equal to the $p_{H_3O}^+$. In ammonia the solvent cation NH_4^+ is a much weaker acid, so that $p_{NH_4}^+$ will be different from the p_H-value, but normal p_H color indicators may be used in most of the proton-containing solvents.

The donor properties of ammonia are stronger than those of water. It is difficult, however, to assign a donor number to a hydrogen-bridging donor solvent because of the unknown influence of hydrogen bonding. Other protonic donor-solvents are hydrazine, carboxylic acids, alcohols and amides, while hydrogen halides, hydrogen cyanide, sulphuric, nitric and phosphoric acids may be considered as acceptor solvents (Chapter IV).

2. Liquid Ammonia

Liquid ammonia is one of the best-known non-aqueous solvents and serves to illustrate some general points. For more detailed information various excellent review articles must be consulted[2-17]. In solid ammonia hydrogen bonding has been established[18] and similar considerations explain the considerable association in the liquid state[16].

According to the highly basic properties heats of solutions in liquid ammonia are frequently higher than in water[19].

[2] FRANKLIN, E. C.: "The Nitrogen System of Compounds," Reinhold Publ. Corp. New York 1935.

[3] YOST, D. M., and H. RUSSELL, Jr.: "Systematic Inorganic Chemistry," Prentice Hall, Englewood Cliffs., N. J. 1946.

[4] JANDER, G.: „Die Chemie in wasserähnlichen Lösungsmitteln", Springer-Verlag, Berlin-Göttingen-Heidelberg 1949.

[5] BERGSTRÖM, F. W., and W. C. FERNELIUS: Chem. Revs. 12, 43 (1933); 20, 413 (1937).

[6] FERNELIUS, W. C., and G. B. BOWMAN: Chem. Revs. 26, 3 (1940).

[7] AUDRIETH, L. F., and J. KLEINBERG: "Non-Aqueous Solvents," Wiley, New York 1953.

[8] KRAUS, C. A.: J. Chem. Educ. 30, 83 (1953).

[9] BIRCH, A. J.: Quart. Revs. 4, 69 (1950).

[10] WATT, G. W.: Chem. Revs. 46, 289, 317 (1950).

[11] JOLLY, W. L.: J. Chem. Educ. 33, 512 (1956).

[12] BIRCH, A. J., and H. SMITH: Quart. Revs. 12, 17 (1958).

[13] SISLER, H. H.: "Chemistry in Non-Aqueous Solvents," Reinhold Publ. Corp. New York 1961.

[14] FOWLES, G. W. A., and D. NICHOLLS: Quart. Revs. 16, 19 (1962).

[15] JOLLY, W. L., and C. J. HALLADA: "Non-Aqueous Solvent Systems," Ed. T. C. WADDINGTON, Acad. Press, London and New York 1965.

[16] JANDER, J.: Vol. I, Part 1 of "Chemistry in Non-Aqueous Ionizing Solvents," Ed. C. C. ADDISON, G. JANDER, and H. SPANDAU, Vieweg and Interscience, Braunschweig, New York and London. 1966.

[17] SMITH, H.: Vol. I, Part 2 of "Chemistry in Non-Aqueous Ionizing Solvents," Ed. G. JANDER, H. SPANDAU, and C. C. ADDISON, Vieweg and Interscience, Braunschweig, New York and London. 1963.

[18] OLAVSSON, J., and D. H. TEMPLETON: Acta Cryst. 12, 840 (1959).

[19] GUNN, S. R., and L. R. GREEN: J. Phys. Chem. 64, 1066 (1960).

Table 14. *Some Physical Properties of Ammonia*

Melting Point (°C)	−77.8
Boiling Point at 760 Torr (°C)	−33.5
Density at b.p. (g · cm⁻³)	0.681
Mol.Vol. at b.p. (cm³ · g⁻¹)	25.0
Specific Conductivity at 33° (Ohm⁻¹ · cm⁻¹)	5.10^{-11}
Dielectric Constant at −33°	23.0
Dipole Moment (Debye)	0.93
Viscosity (Centipoise) at b.p.	0.2543
Heat of Vaporisation (kcal · mole⁻¹)	5.58

Table 15. *Heats of Solution in Water and Ammonia*

Compound	ΔH in H_2O (25°)	ΔH in NH_3 (−33°)
NaCl	+1.02	−11.57
KI	+4.95	− 7.89
NH₄Cl	+3.71	− 6.95
NH₄I	+3.3	−13.37

That ammonia is a stronger donor solvent than water can also be seen from the ease of formation of certain ammoniated transition metal ions in aqueous solutions. Numerous crystalline ammoniated compounds are known, in which the nitrogen atom with its free electron pair acts as the donor atom; in numerous compounds hydrogen bridging is known to occur and proton transfer reactions are easily accomplished.

$$NH_4^+ + NH_3 \rightleftharpoons NH_3 + NH_4^+$$

While in water the H_3O^+-ion shows an abnormal proton jump conduction mechanism, the equivalent ion in liquid ammonia, NH_4^+, exhibits no such abnormality and this is attributed to the fact that it is completely coordinated and possesses no dipole moment[1].

Due to the high donor properties many ionic compounds frequently show even higher solubilities in liquid ammonia than in water. Some values have been tabulated recently by JANDER[16] and a few examples are given in Table 16.

Table 16. *Solubilities of Some Alkali Salts in g/100 g Solvent*

Compound	Solubility in H_2O at 20°	Solubility in NH_3 at −35°
LiNO₃	52	243
NaNO₃	92	98
KNO₃	38	10
NaCl	37	3

Nitrates, cyanides, thiocyanates are easily soluble as are many other 1 : 1 electrolytes. The solubilities of halides increase usually from fluoride to iodide, the solubilities of silver halides showing the reversed order from that in water.

On the other hand 1 : 2 electrolytes and polyelectrolytes are frequently less soluble. Examples are numerous carbonates, sulphates, phosphates and oxides. Halides and pseudohalides of the alkaline earth metals, which belong also to this group, show considerably smaller solubilities than the corresponding alkali metal compounds.

Due to the strong interaction between ammonia and certain transition metal ions, such as Co^{2+}, Ni^{2+}, Cu^{2+}, Zn^{2+} and Ag^+, salts of such metal ions show high solubilities in liquid ammonia.

Compounds capable of forming hydrogen bonds with ammonia show high solubilities in this solvent, just as in other hydrogen bonding solvents. Examples are carbohydrates, esters, amines and phenols.

Numerous metathetical reactions are possible, which cannot be carried out in water. A well known example is the reaction of AgCl with $Ba(NO_3)_2$ which yields nearly insoluble $BaCl_2$:

$$Ba(NO_3)_2 + 2\ AgCl \rightarrow BaCl_2\downarrow + 2\ AgNO_3$$

Many amides of heavy metal ions are readily prepared due to their poor solubilities in liquid ammonia according to the reactions[20]:

$$AgNO_3 + KNH_2 \rightarrow AgNH_2 + KNO_3$$
$$Pb(NO_3)_2 + 2\ KNH_2 \rightarrow Pb(NH_2)_2 + 2\ KNO_3$$
$$ZnI_2 + 2\ KNH_2 \rightarrow Zn(NH_2)_2 + 2\ KI$$

Some of them have higher solubilities in the presence of large excess of alkali amides[21–24] and thus show amphoteric properties in liquid ammonia:

$$AgNH_2 + NH_2^- \rightleftharpoons [Ag(NH_2)_2]^-$$
$$Zn(NH_2)_2 + 2\ NH_2^- \rightleftharpoons [Zn(NH_2)_4]^{2-}$$

Since sulphides of the alkaline earth metals show small solubilities, they too can be precipitated according to:

$$(NH_4)_2S + Ca(NO_3)_2 \rightarrow CaS\downarrow + 2\ NH_4NO_3$$
$$(NH_4)_2S + Ba(NO_3)_2 \rightarrow BaS\downarrow + 2\ NH_4NO_3$$

Liquid ammonia is associated through hydrogen bonds, the bond energies of which are somewhat smaller than those in water. The small conductivity of the purified solvent has been attributed to self-ionization of the liquid and this has been confirmed by potentiometric measurements[16, 25]:

$$\mathrm{H^+}$$
$$NH_3 + NH_3 \rightleftharpoons NH_4^+ + NH_2^-$$

Due to the high proton acceptor properties of ammonia nearly any potential proton donor will be able to act as an acid, allowing liquid ammonia to be a

[20] AUDRIETH, L. F.: Angew. Chem. 45, 385 (1934).

[21] FRANKLIN, E. C.: J. Phys. Chem. 15, 509 (1901).

[22] BERGSTRÖM, F. W.: J. Amer. Chem. Soc. 45, 2792 (1923); 46, 1545 (1924); 48, 2848 (1926); 50, 652 (1928).

[23] FITZGERALD, F. F.: J. Amer. Chem. Soc. 29, 656 (1907).

[24] FRANKLIN, E. C.: J. Amer. Chem. Soc. 29, 274 (1907).

[25] PLESKOV, V. A., and A. M. MONOSZON: Acta Phys.-Chim. USSR 1, 713; 2, 615 (1935).

levelling solvent for acids; it is true that proton transfer is more complete for acetic acid in liquid ammonia but it would be erroneous to assume from its enhanced ionization that the acidic properties should also be increased. In fact the dissociation constants of acetic acid in water and of ammonium acetate in liquid ammonia are about equal (Table 17). It must be remembered that the NH_4^+ ion is a weaker acid in liquid ammonia than is H_3O^+ in water, and that K_a as defined by BRÖNSTED is merely a measure of the strength of the acid in a given solvent. Moreover, we have already said that it is without meaning to compare the dissociation constants of an acid in different solvents[26]. Since ammonium sulphate is scarcely soluble, the acidity of sulphuric acid solutions in liquid ammonia is small.

Table 17. *Dissociation Constants of "Ammono Acids" in Liquid Ammonia at* $-33°$

Acid	K_a in NH_3
NH_4ClO_4	$5.4 \cdot 10^{-3}$
NH_4NO_3	$4.3 \cdot 10^{-3}$
NH_4Cl	$1.3 \cdot 10^{-3}$
NH_4Br	$2.4 \cdot 10^{-3}$
NH_4CN	$1.9 \cdot 10^{-3}$
$(NH_4)_2S$	$9.8 \cdot 10^{-4}$
NH_4OOCCH_3	$0.7 \cdot 10^{-4}$

The rate at which protons are exchanged in liquid ammonia has been measured by n.m.r. techniques[27]. Three peaks appear in the proton magnetic resonance spectrum of pure liquid ammonia due to spin-spin coupling of the protons with the ^{14}N-nucleus. When a trace of ammonium salt is added the signal changes to a single line, which is attributed to

$$NH_3 \; + \; NH_4^+ \rightleftharpoons NH_4^+ + NH_3$$

From a rough estimate of the ammonium ion concentration required to cause collapse of the triplet, the rate constant was estimated to be 5.10^8 liter $\cdot$ mole^{-1} sec^{-1}.

The exchange between solvent molecules and those coordinated to the ions $[Ag(NH_3)_2]^+$, $[Cu(NH_3)_4]^{2+}$ and $[Ni(NH_3)_6]^{2+}$ is very fast[28], but a bimolecular rate constant of $1.3 \cdot 10^{-6}$ was found for the exchange of $[Cr(NH_3)_6]^{3+}$. A unimolecular reaction with a rate constant of $k = 4.7 \cdot 10^4$ sec^{-1} at $25°$ was proposed from the rate of exchange of ^{14}N between $[Ni(NH_3)_6]^{2+}$ and the solvent ammonia molecules by n.m.r. line broadening measurements[28].

The ionic product of the self-ionization of liquid ammonia has been determined from potentiometric measurements using the cell[25]

$$Pt, H_2 / 0.1 \text{ M } NH_4NO_3 / \text{sat. } KNO_3 / KNH_2 / Pt, H_2$$

[26] MANDEL, M.: Nature **176**, 792 (1955).

[27] OGG, R. A., Jr.: Disc. Faraday Soc. No. **17**, 215 (1954).

[28] WIESENDANGER, H. U. D., W. H. JONES, and C. S. GARNET: J. Chem. Phys. **27**, 668 (1937).

to be $K = 1.9 \cdot 10^{-33}$ at $-50°$, while figures ranging from $K = 5.10^{-27}$ to $K = 2.20^{-28}$ are derived from other measurements[29, 30].

In liquid ammonia soluble amides act as bases, which may formally be compared with hydroxides in water. Likewise imides such as Pb(NH) are formally analogous to oxides in water which by addition of the solvent may display basic properties. In liquid ammonia a third type of potential base is known, namely nitrides.

In water:
$$NaOH \rightleftharpoons Na^+ + OH^-$$
$$PbO + H_2O \rightleftharpoons Pb^{++} + 2\,OH^-$$

In liquid NH_3:
$$NaNH_2 \rightleftharpoons Na^+ + NH_2^-$$
$$Pb(NH) + NH_3 \rightleftharpoons Pb^{++} + 2\,NH_2^-$$
$$Mg_3N_2 + 4\,NH_3 \rightleftharpoons 3\,Mg^{++} + 6\,NH_2^-$$

Much of the coordination chemistry in liquid ammonia is due to "solvolytic reactions," which in liquid ammonia are frequently called "ammonolytic reactions" or "ammonolysis." It is impossible to give a full account of this type of reactions in this presentation, but a few examples may be mentioned:

Phosphorus oxychloride is hydrolysed by water to give phosphoric acid and ammonolysis leads to phosphorus oxytriamide[31]:

$$POCl_3 + 3\,H_2O \rightarrow PO(OH)_3 + 3\,HCl$$
$$POCl_3 + 6\,NH_3 \rightarrow PO(NH_2)_3 + 3\,NH_4Cl$$

Phosphorus(V) chloride gives with liquid ammonia a series of phosphonitrilic compounds:

$$n\,PCl_5 + n\,NH_3 \rightarrow (PNCl_2)_n + 3\,n\,HCl$$

of which crystalline ring compounds and amorphous chains are known[32]:

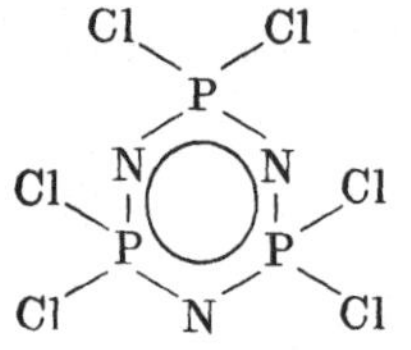

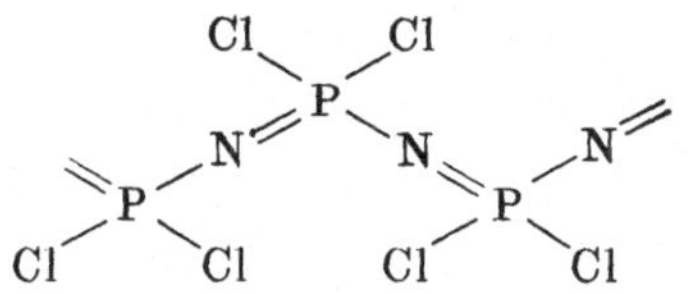

Ammonolysis of $(C_6H_5NPCl_3)_2$ occurs according to[33]:

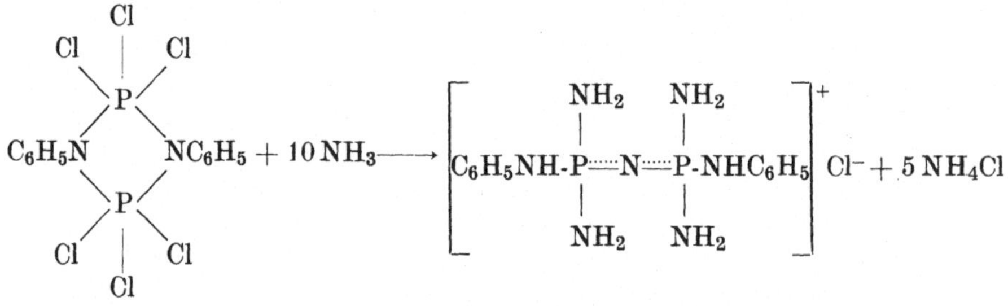

[29] MULDER, H. D., and F. C. SCHMIDT: J. Amer. Chem. Soc. **73**, 5575 (1951).

[30] COULTER, L. V., J. R. SINCLAIR, A. G. COLE, and G. C. ROPER: J. Amer. Chem. Soc. **81**, 2986 (1959).

[31] KLEMENT, R., and O. KOCH: Chem. Ber. **87**, 333 (1954).

[32] BECKE-GOEHRING, M., and E. FLUCK: Angew. Chem. **74**, 382 (1962).

[33] UTVARY, K., V. GUTMANN, and CH. KEMENATER: Mh. Chem. **96**, 1751 (1965).

With sulphur chlorides tetrasulphurtetranitride is readily produced by disproportionation:

$$6\,S_2Cl_2 + 16\,NH_3 \rightarrow S_4N_4 + 12\,NH_4Cl + 8\,S$$

Silicon(IV) fluoride gives a hexacoordinated coordination compound[34, 35]:

$$F - Si - F + 2\,NH_3 \longrightarrow$$

but silicon tetrachloride is ammonolysed to give polymeric silicon diimide[36]:

$$n\,SiCl_4 + 6\,n\,NH_3 \rightarrow [Si(NH)_2]_n + 4\,n\,NH_4Cl$$

Likewise boron(III) fluoride gives the coordination compound $BF_3 \cdot NH_3$, but boron(III) chloride is ammonolysed[37]:

$$BF_3 + NH_3 \rightarrow H_3N : BF_3$$
$$BCl_3 + 3\,NH_3 \rightarrow B(NH_2)_3$$

Diborane gives a conducting solution in liquid ammonia, which was thought to contain $[NH_4]^+[BH_3-NH_2-BH_3]^-$, but was later shown to contain the following species[38]:

$$\left[\begin{array}{cc} H & NH_3 \\ & B \\ H & NH_3 \end{array}\right]^+ \quad \text{and} \quad \left[\begin{array}{cc} H & H \\ & B \\ H & H \end{array}\right]^-$$

Reactions of metal halides with ammonia may involve the formation of addition compounds or bring about ammonolysis of one or more of the metal-halogen bonds[39]. The dichlorides of manganese, iron, cobalt and nickel form adducts. Vanadium(III) chloride and vanadium(III) bromide are ammonolysed in liquid ammonia[40, 41] to a mixture of ammonolysis products with the same chemical composition as that of the previously reported hexamine:

$$VX_3 + 6\,NH_3 \rightarrow VX_2(NH_2) \cdot 4\,NH_3 + NH_4X$$

Likewise polymeric molybdenum(III) bromide is ammonolysed[42].

When titanium(IV) chloride is treated with liquid ammonia, a series of equilibria appear to be involved[39]

$$TiCl_4 + 2\,NH_3 \rightleftharpoons TiCl_3(NH_2) + NH_4Cl$$
$$TiCl_3(NH_2) + 2\,NH_3 \rightleftharpoons TiCl_2(NH_2)_2 + NH_4Cl$$
$$TiCl_2(NH_2)_2 + 2\,NH_3 \rightleftharpoons TiCl(NH_2)_3 + NH_4Cl$$
$$TiCl(NH_2)_3 + 2\,NH_3 \rightleftharpoons Ti(NH_2)_4 + NH_4Cl$$

[34] PIPER, T. S., and E. G. ROCHOW: J. Amer. Chem. Soc. **76**, 4318 (1954).
[35] MUETTERTIES, E. L.: J. Amer. Chem. Soc. **82**, 1082 (1960).
[36] SCHENK, P. W., and J. B. P. TRIPATHI: Angew. Chem. **74**, 116 (1962).
[37] JOANNIS, A.: Compt. Rend. Acad. Sci. Paris **135**, 1106 (1902).
[38] PARRY, R. W., D. R. SCHULTZ, and P. R. GIRARDOT: J. Amer. Chem. Soc. **80**, 1 (1958).
[39] FOWLES, G. W. A.: Progr. Inorg. Chem. **6**, 1 (1964).
[40] FOWLES, G. W. A., P. G. LANIGAN, and D. NICHOLLS: Chem. Ind. (London) **1961**, 1167.
[41] NICHOLLS, D.: J. Inorg. Nucl. Chem. **24**, 1001 (1962).
[42] EDWARDS, D. A., and G. W. A. FOWLES: J. Less Common Metals **4**, 512 (1962).

Moreover polymerization may occur leading to insoluble products in which the halogen to metal ratio is non-integral and for which various structures were proposed:

$$
M\overset{\displaystyle X}{\underset{\displaystyle X}{\diamond}}M; \quad M\overset{\displaystyle NH_2}{\underset{\displaystyle NH_2}{\diamond}}M; \quad M\overset{\displaystyle NH}{\underset{\displaystyle NH}{\diamond}}M; \quad M\overset{\displaystyle NH-H\ldots X}{\underset{\displaystyle X\ldots H-HN}{\diamond}}M
$$

When potassium amide is added to the ammonolysed solution of titanium(IV) chloride Ti(NH)(NK) is produced[39].

The complex salts M_2TiX_6 give more soluble ammonolysis products than do the corresponding tetrahalides[43]. When the potassium salt is used, potassium chloride is precipitated:

$$
K_2TiCl_6 \xrightarrow{\;NH_3\;} K_2[TiCl_4(NH_2)_2] + 2\,NH_4Cl
$$

$$
\Big\downarrow NH_3
$$

$$
(NH_4)_2[TiCl_4(NH_2)_2] + 2\,KCl
$$

and on further reaction with liquid ammonia ammonolysis is continued to give $TiCl_2(NH_2)_2$.

Zirconium(IV) chloride is more resistant towards ammonolysis presumably because the zirconium-chlorine bonds are strong[39]. The insoluble product of approximate composition $ZrCl_3(NH_2)$ is probably polymeric. No ammonolysis has been observed with thorium(IV) halides.

Vanadium(V) fluoride is reduced by ammonia to give the adduct $VF_4 \cdot NH_3$[44]. The highest fluorides of Nb, Ta, Mo and W are resistant towards ammonolysis, while the chlorides are not[45].

$$
VCl_4 + 6\,NH_3 \rightarrow VCl(NH_2)_3 + 3\,NH_4Cl
$$

The pentachlorides of niobium and tantalum dissolve completely in liquid ammonia[46, 47] so that separation from ammonium chloride is difficult, but tensimetric experiments suggest solvolysis according to:

$$
NbCl_5 + 4\,NH_3 \rightarrow NbCl_3(NH_2)_2 + 2\,NH_4Cl
$$

With $NbBr_5$ an ammonobasic compound of composition $NbBr(NH_2)_2(NH)$ is precipitated.

Two Mo-Cl bonds of molybdenum(V) chloride are ammonolysed at $-33°$ and four Mo-Cl bonds at room temperature to give $MoCl(NH)(NH_2)_2$, which is scarcely soluble and probably polymeric.

Polymeric species have also been obtained by ammonolysing chromium(III) and cobalt(III) compounds by alkali amides[48] where the first step may be represent-

[43] FOWLES, G. W. A., and D. NICHOLLS: J. Chem. Soc. **1961**, 65.

[44] CAVELL, R. G., and H. C. CLARK: J. Inorg. Nucl. Chem. **17**, 257 (1961).

[45] FOWLES, G. W. A., and D. NICHOLLS: J. Chem. Soc. **1958**, 1687.

[46] FOWLES, G. W. A., and F. H. POLLARD: J. Chem. Soc. **1952**, 4938.

[47] MOUREN, H., and C. HAMBLETT: J. Amer. Chem. Soc. **59**, 33 (1937).

[48] SCHMITZ-DUMONT, O., J. PILZECKER, and H. F. PIEPENBRINK: Z. anorg. allg. Chem. **248**, 175 (1941).

ed by the equation:

$$2\left[Co(NH_3)_6\right]^{3+} \longrightarrow 2\,NH_2^- + \left[(H_3N)_4Co\begin{matrix} H & H \\ \diagdown N \diagup \\ \diagup \quad \diagdown \\ \diagdown \quad \diagup \\ \diagup N \diagdown \\ H & H \end{matrix}Co(NH_3)_4\right]^{4+}$$

The final product has been formulated as

$$(H_3N)_3Co\begin{matrix}\diagup NH_2\diagdown \\ -NH_2- \\ \diagdown NH_2\diagup\end{matrix}\left[\begin{matrix}\diagup NH_2\diagdown \\ Co-NH_2- \\ \diagdown NH_2\diagup\end{matrix}\right]_n Co(NH_3)_3$$

but a three-dimensional network seems more likely. Likewise highly polymeric titanium(III) amide has been described[49]:

$$K_3[Ti(SCN)_6] + 3\,KNH_2 \xrightarrow{\text{liq.NH}_3} Ti(NH_2)_3 + 6\,KSCN$$

Liquid ammonia is unique in its ability to dissolve alkali and alkaline earth metals to give blue solutions. Such solutions are metastable. After standing for long periods in the presence of suitable catalysts the reaction follows the pattern known for aqueous systems with liberation of hydrogen.

$$Na + NH_3 \xrightarrow{\text{catalyst}} NaNH_2 + \tfrac{1}{2}H_2$$

In the absence of a catalyst the solutions show much higher conductivities than those of any other electrolyte in any solvent owing to the presence of electrons, which are more or less solvated[50-52]:

$$M = M^+ + e^-$$

The specific conductivities of concentrated solutions are by one order of magnitude higher than those of salts in water[51, 53-57]. With increasing concentration the conductivity decreases and passes through a minimum for 0.04 M solutions. Transport number measurements on sodium solutions have shown that the equivalent conductance of the anion (the solvated electron) has a minimum value at 0.04 M, while the equivalent conductance of the metal ion decreases continuously with increase in concentration[58]. The blue colour of dilute metal solutions is due to the short wave length tail of a broad absorption band with its peak at 15,000 Å. The

[49] SCHMITZ-DUMONT, O., P. SIMONS, and G. BROJA: Z. anorg. allg. Chem. 258, 307 (1949).
[50] KRAUS, C. A.: J. Amer. Chem. Soc. 30, 653, 1197, 1323 (1908).
[51] KRAUS, C. A.: J. Amer. Chem. Soc. 43, 749 (1921).
[52] KRAUS, C. A.: Chem. Revs. 8, 251 (1931).
[53] CADY, H. P.: J. Phys. Chem. 1, 707 (1897).
[54] KRAUS, C. A.: J. Amer. Chem. Soc. 44, 1941 (1922).
[55] GIBSON, G. E., G. E. PHIPPS, and T. E. PHIPPS: J. Amer. Chem. Soc. 48, 312 (1926).
[56] JOHNSON, W. C., and A. W. MEYER: Chem. Revs. 8, 273 (1931).
[57] KRAUS, C. A., and W. W. LUCASSE: J. Amer. Chem. Soc. 43, 2529 (1921).
[58] DYE, J. L., R. F. SANKUER, and G. E. SMITH: J. Amer. Chem. Soc. 82, 4797 (1960).

spectra of the alkali metals and of the alkaline earth metals (at least calcium) are identical[59, 60].

The molar magnetic susceptibility of an alkali metal solution approaches that of one mole of free electron spins at infinite dilution; it is decreased by increase in concentration[61]. Due to the magnetic data the model given by KRAUS is not adequate since it does not account for the pairing of electrons in the solutions. The following equilibria have been suggested with the respective equilibrium constants:

$$M \ = M^+ + e^-; \qquad k_1 = 9.9 \cdot 10^{-3}$$
$$M^- = M + e^-; \qquad k_2 = 9.7 \cdot 10^{-4}$$
$$M_2 = 2\,M; \qquad k_3 = 1.9 \cdot 10^{-4}$$

Numerous reactions have been carried out with metal-ammonia solutions, which have strongly reducing properties. They have been extensively reviewed[9,11,12, 14,15] so that only a few examples which might be of particular interest to coordination chemistry, will be quoted: oxygen, sulphur, selenium and tellurium give a number of isopolyanionic compounds by simple electron addition to the elements[62-67]:

$$O_2 + e^- \rightarrow O_2^-$$

Many compounds are reduced to the free elements, to intermetallic compounds or to homopolyatomic anionic complexes containing the reduced elements. For example with lead iodide the compound $Na_4[Pb(Pb)_8]$ has been obtained.

The formation of compounds containing the elements with less familiar oxidation states is another feature of chemistry in liquid ammonia solution[68]. For example tetracyanonickelate(II) is reduced by potassium in liquid ammonia at $-33°$ to the red cyanonickelate(I)[69], which is slowly reduced further at $0°$ to the yellow cyanonicklate(0) $K_4[Ni(CN)_4]$[70]:

$$K_2[Ni(CN)_4] + K \xrightarrow{\text{liq. NH}_3} K_3[Ni(CN)_4]$$
$$K_3[Ni(CN)_4] + K \xrightarrow{\text{liq.NH}_3} K_4[Ni(CN)_4]$$

No cyanonickelate(I) is, however, obtained when cyanonickelate(II) is reduced by excess of potassium[70]. Similarly the compounds $K_4[Co(CN)_4]$[71] and $K_4[Pd(CN)_4]$[72] are produced, which contain the transition metals in the zero oxidation states. On the other hand the reduction of the cyanocomplex of chro-

[59] GOLD, M., and W. L. JOLLY: Inorg. Chem. **1**, 818 (1962).

[60] HALLADA, C. J., and W. L. JOLLY: Inorg. Chem. **2**, 1076 (1963).

[61] FREED, S., and N. SUGARMANN: J. Chem. Phys. **11**, 354 (1943).

[62] ZINTL, E., J. GOUBEAU, and W. DULLENKOPF: Z. phys. Chem. A **154**, 1 (1931).

[63] ZINTL, E., and G. A. HARDER: Z. physik. Chem. A **154**, 47 (1931).

[64] SCHECHTER, W. H., H. H. SISLER, and J. KLEINBERG: J. Amer. Chem. Soc. **70**, 267 (1948).

[65] SCHECHTER, W. H., J. K. THOMPSON, and J. KLEINBERG: J. Amer. Chem. Soc. **71**, 1816 (1949).

[66] THOMPSON, J. K., and J. KLEINBERG: J. Amer. Chem. Soc. **73**, 1243 (1951).

[67] STEPHANOU, S. E., W. H. SCHECHTER, W. J. ARGERSINGER, Jr., and J. KLEINBERG: J. Amer. Chem. Soc. **71**, 1819 (1949).

[68] COLTON, E.: J. Chem. Educ. **31**, 527 (1954).

[69] EASTES, J. W., and W. M. BURGESS: J. Amer. Chem. Soc. **64**, 1187 (1942).

[70] WATT, G. W., J. L. HALL, G. R. CHOPPIN, and P. S. GENTILE: J. Amer. Chem. Soc. **76**, 373 (1954).

[71] HIEBER, W., and C. BARTENSTEIN: Naturwiss. **13**, 300 (1952).

[72] BURBAGE, J. J., and W. C. FERNELIUS: J. Amer. Chem. Soc. **65**, 1484 (1943).

mium(III) gives a product containing chromium(I) and that of the corresponding compound of manganese yields a mixture of complex cyanides containing manganese(I) and manganese(0)[73].

The reduction of bromopentammin-iridium(III) bromide leads to the interesting compound $[Ir(NH_3)_4]$[74], which is insoluble in liquid ammonia. There is also evidence for a less stable ammine of platinum(0)[75], as well as for an ethylene-diamine compound.

$$[Pt(NH_3)_4]^{2+} + 2\,e^- \rightarrow [Pt(NH_3)_4]$$
$$[Pt(en)_2]^{2+} + 2\,e^- \rightarrow [Pt(en)_2]$$

Carbonyls may be reduced by alkali metal solutions in liquid ammonia to give carbonyl metallates. These are salt-like compounds containing the metals with negative oxidation numbers in the complex anions:

$$Mn_2(CO)_{10} + 2\,K \rightarrow 2\,K^+ + 2\,[Mn(CO)_5]^-$$
$$Fe(CO)_5 + 2\,Na \rightarrow 2\,Na^+ + [Fe(CO)_4]^{2-} + CO$$
$$Co_2(CO)_8 + 2\,Na \rightarrow 2\,Na^+ + 2\,[Co(CO)_4]^-$$

Another interesting example is the reduction of nitrogen oxide[76], which gives a compound probably containing the nitroside ion NO^-:

$$Na + NO \rightarrow Na^+NO^-$$

Many organic compounds are reduced by metal-ammonia solutions. Acetylenes are converted to acetylide ions:

$$RC \equiv CH + e^- \rightarrow RC \equiv C^- + H$$

which may be further reduced to ethylene derivatives:

$$RC \equiv CH + 2\,H \rightarrow RCH = CH_2$$

From alkyl halides the halide ion may be removed by metal ammonia solutions

$$RX + 2\,e^- \rightarrow R^- + X^-$$

in some cases leading to the corresponding hydrocarbons:

$$R^- + NH_3 \rightarrow RH + NH_2^-$$

Alcohols and phenols form readily alkoxides and phenoxides. Organometallic chemistry is making use of metal ammonia solutions *inter alia* to prepare compounds containing metal-metal bonds:

$$(C_6H_5)_3GeCl + 2\,Na \xrightarrow{\text{liq. } NH_3} (C_6H_5)_3GeNa + NaCl$$
$$(C_6H_5)_3GeNa + BrSn(CH_3)_3 \rightarrow (C_6H_5)_3Ge-Sn(CH_3)_3 + NaBr$$

Many other redox reactions have been observed in the absence of metals dissolved in liquid ammonia[16]. Some compounds are reduced by the solvent, for example[77]

$$3\,TeCl_2 + 8\,NH_3 \rightarrow 3\,Te + 6\,NH_4Cl + N_2$$

and others undergo disproportionation reactions[78]:

$$3\,[Co^0(CO)_4]_2 + 12\,NH_3 \rightarrow 2\,[Co^{II}(NH_3)_6][Co^{-1}(CO)_4]_2 + 8\,CO$$

[73] DAVIDSON, W. A., and J. KLEINBERG: J. Phys. Chem. **57**, 571 (1953).
[74] WATT, G. W., and P. J. MAYFIELD: J. Amer. Chem. Soc. **75**, 6178 (1953).
[75] WATT, G. W., M. T. WALLING, and P. J. MAYFIELD: J. Amer. Chem. Soc. **75**, 6175 (1953).
[76] FRAZER, J. H., and N. LONG: J. Chem. Phys. **6**, 462 (1938).
[77] AYNSLEY, E. E.: J. Chem. Soc. **1953**, 3016.
[78] BEHRENS, H., and R. WEBER: Z. anorg. allg. Chem. **281**, 190 (1955).

3. Hydrazine

Table 18. *Some Physical Properties of Hydrazine*

Melting Point (°C)	+2
Boiling Point (°C)	+113.5
Density at 35° (g · cm^{-3})	0.9955
Specific Conductivity at 0° (Ohm^{-1} · cm^{-1})	1.10^{-6}
Dielectric Constant (25°)	51.7
Heat of Vaporization (kcal · mole^{-1})	10.2

$$\begin{array}{c} H\diagdown \\ \\ H\diagup \end{array} N \xrightarrow{1.47} N \begin{array}{c} \diagup H \\ \\ \diagdown H \end{array}$$
$$108°$$

Hydrazine is in the liquid state associated[79] through hydrogen bonding and is monomolecular in the vapour phase[80]. It forms an azeotropic mixture with water, which has approximately the same boiling point as anhydrous hydrazine. Barium oxide is useful as a dehydrating agent and the anhydrous solvent may also be obtained by reacting hydrazinium salts with liquid ammonia[81,82] due to the in-solubility of ammonium sulphate in this solvent.

$$(N_2H_5)_2SO_4 + 2NH_3 \rightleftarrows 2N_2H_4 + (NH_4)_2SO_4\downarrow$$

In water due to the higher solubility of the sulphate this equilibrium is shifted to the opposite side.

In water: $2N_2H_4 + (NH_4)_2SO_4 \rightleftarrows (N_2H_5)_2SO_4 + 2NH_3\uparrow$

Hydrazine is a weakly endothermic compound, which is easily decomposed. For storage certain metal salts, such as sulphides or thiocyanates, are used as stabilizing agents.

Many ionic compounds are soluble in anhydrous hydrazine and give conducting solutions. Alkali halides, tetraalkylammonium halides and many carboxylic acids are practically completely dissociated due to the reasonably high dielectric constant of the solvent. Various covalent organic compounds such as a variety of nitrocompounds which show poor solubilities in water, give conducting solutions in anhydrous hydrazine, for which the following dissociation is assumed[83]:

$$C_6H_5NO_2 + N_2H_4 \rightleftharpoons N_2H_4^+ + C_6H_5NO_2^-$$

On the other hand many oxides, phosphates, sulphates and carbonates are nearly insoluble[7]. Low solubilities are also found for ZnS, Sb$_2$S$_5$, LaCl$_3$, KMnO$_4$ and halides of many transition metals, such as AgCl or PtCl$_4$.

Another feature of liquid hydrazine is exhibited by its reducing properties[7,84]. Thus halides of silver, copper(II), mercury(II), arsenic(III), bismuth(III), platinum(II) and palladium(II) are reduced to the respective metals.

[79] WALDEN, P., and H. HILGERT: Z. Phys. Chem. **A 165**, 241 (1933).
[80] GIGUÈRE, P. A., and R. E. RUNDLE: J. Amer. Chem. Soc. **63**, 1135 (1941).
[81] BROWNE, A. W., and T. W. B. WELSH: J. Amer. Chem. Soc. **33**, 1728 (1911).
[82] AUDRIETH, L. F., and P. H. MOHR: Ind. Eng. Chem. **43**, 1774 (1951).
[83] WALDEN, P., and H. HILGERT: Z. Phys. Chem. **A 168**, 419 (1934).
[84] AUDRIETH, L. F., and P. H. MOHR: Chem. Eng. News **26**, 3746 (1948).

The following solvates have been described[85]: $ZnCl_2(N_2H_4)_2$, $InCl_3(N_2H_4)_3$, $InBr_3(N_2H_4)_3$, $NiSO_4(N_2H_4)_3$ and $Co(SCN)_2(N_2H_4)_3$. No crystal structures are available, although they would be of interest in establishing the coordinating properties of hydrazine, which might behave as a bidentate coordinating group.

Most of the known chemistry in liquid hydrazine covers protolytic reactions, and may be referred to the self-ionization in the pure liquid state[7,86]:

$$2 N_2H_4 \rightleftharpoons N_2H_5^+ + N_2H_3^-$$

Hydrazinium compounds such as hydrazinium perchlorate[87] act as acids in this medium. Apart from the $N_2H_5^+$-ion the existence of $N_2H_6^{2+}$ has been shown in compounds like hydrazinium difluoride $N_2H_6F_2$[88].

Since the proton affinity of hydrazine is similar to that of ammonia, acid-base reactions are similar to those in this solvent system. Many compounds are known to act as moderately strong acids, but only a few substances (notably the hydrazides of the alkali metals) behave as bases. NaN_2H_3 may be obtained by dropwise addition of hydrazine to a suspension of finely divided sodium in dry ether. Electrolysis in liquid hydrazine gives nitrogen at the anode and hydrogen at the cathode[89].

Table 19. *Conductivities in Hydrazine*

Electrolyte	λ_0	Electrolyte	λ_0
$(C_2H_5)_4NCl$	100.6	Nitromethane	88.0
$(C_2H_5)_4NBr$	105.3	o-Nitroaniline	87.0
$(C_2H_5)_4NI$	102.9	o-Dinitrobenzene	170.0
$(C_3H_7)_4NClO_4$	93.0	1,3,5-Trinitrobenzene	277.0
N_2H_5Cl	156.5	Picric Acid	269.0
$LiClO_4$	69.6	Benzoic Acid	86.5
$NaClO_4$	111.4	Triphenylacetic Acid	75.0
$KClO_4$	91.5	o-Phthalic Acid	206.5
KI	132.9	Maleic Acid	215.0
KCl	130.3	Fumaric Acid	215.7
NaI	114.5	Terephthalic Acid	206.5
CdI_2	77.2		

4. Hydrogen Sulphide

Hydrogen sulphide is a poor ionizing solvent of low donor properties. It is only very weakly hydrogen bonded in the liquid state and has a low dielectric constant[4]; the donor properties are considerably lower than those of water and indeed acceptor properties are also apparent.

Autoprotolysis has been postulated according to

$$2 H_2S \rightleftharpoons H_3S^+ + SH^-$$

[85] SUTTON, G. J.: Australian Chem. Inst. J. and Proc. **16**, 115 (1949); Chem. Abstr. **43**, 6932 i (1949).

[86] AUDRIETH, L. F.: Z. Physik. Chem. **A 165**, 323 (1933).

[87] BARLOT, J., and S. MARSAULE: Compt. Rend. Acad. Sci. Paris **228**, 1497 (1949).

[88] KRONBERG, M. L., and D. HARKER: J. Chem. Phys. **10**, 309 (1942).

[89] WELSH, T. W. B.: J. Amer. Chem. Soc. **37**, 497 (1915).

The autoprotolysis constant was roughly estimated from potentiometric work of little precision[4]

$$K_{H_2S} \sim 10^{-30} \text{ to } 10^{-40}$$

Table 20. *Some Physical Properties of Hydrogen Sulphide*

Melting Point (°C)	-85.5
Boiling Point (°C)	-60.4
Density at $-60°$ (g · cm^{-3})	0.950
Dielectric Constant at $-60°$	10.2
Specific Conductivity at $-78°$ (Ohm^{-1}cm^{-1}) ...	3.10^{-11}
Heat of Vaporization (kcal · mole^{-1})	4.5

Ionic compounds are usually scarcely soluble, but covalent compounds such as hydrogen halides and many organic compounds are soluble. Examples are hydrocarbons, carboxylic acids, acid chlorides, nitriles, aldehydes, ketones, alcohols, amines and nitrocompounds, but the conductivities of their solutions are very small. Many acceptor halides such as those of aluminium, titanium(IV), tin(IV), phosphorus(V), arsenic(III) and bismuth(III) as well as ferric chloride show reasonable solubilities due to interaction with liquid hydrogen sulphide. Solvates of $AlCl_3$, $AlBr_3$, $TiCl_4$, $SnCl_4$, BCl_3 and others have been described[90,91].

Free chlorine and bromine cause oxidation of the solvent, but iodine is soluble possibly with slight electrolytic dissociation:

$$I_2 \rightleftharpoons I^+ + I^-$$

Hydrogen sulphides of the alkali metals, of the ammonium ion and of the tetraalkylsubstituted ammonium ions[92] act as bases in liquid hydrogen sulphide. The SH^- ion is comparable in size with the bromide ion; the hydrosulphides of sodium, potassium and rubidium are scarcely soluble, but suspensions in liquid hydrogen sulphide can be made to react with acids showing moderate ionization of the dissolved hydrogen sulphides.

Since hydrogen sulphide has only a moderate proton affinity, various proton acceptors will behave as bases, for example, triethylamine or pyridine.

$$(C_2H_5)_3N + H_2S \rightleftharpoons (C_2H_5)_3NH^+ + HS^-$$

Hydrogen chloride or sulphuric acid show acidic properties in liquid hydrogen sulphide, but their conductivities are very low.

Neutralization reactions have been reported, e.g.:

$$HCl + NaSH \rightleftharpoons NaCl + H_2S$$

and some of them have been followed by conductometric techniques. Reactions of amines have also been shown to occur. Colour indicators give reversible colour changes and thus may be used to follow neutralization reactions.

Solvolytic reactions can lead to insoluble sulphides[91]:

$$2\,AgCl + H_2S \rightarrow Ag_2S\downarrow + 2\,HCl$$
$$2\,CuCl + H_2S \rightarrow Cu_2S\downarrow + 2\,HCl$$

[90] BILTZ, W., and E. KEUNECKE: Z. anorg. allg. Chem. **147**, 171 (1925).
[91] RALSTON, A. W., and J. A. WILKINSON: J. Amer. Chem. Soc. **50**, 258 (1928).
[92] JANDER, G., and H. SCHMIDT: Wiener Chem. Ztg. **46**, 49 (1943).

The latter reaction yields actually $Cu_{9,5}S_4$[93]. The reaction

$$SnCl_4 + 2H_2S \rightarrow SnS_2\downarrow + 4HCl$$

proceeds at room temperature. Phosphorus(V) chloride and antimony(V) chloride yield the corresponding thiochlorides:

$$PCl_5 + H_2S \rightarrow PSCl_3 + 2HCl$$
$$SbCl_5 + H_2S \rightarrow SbSCl_3 + 2HCl$$

On the other hand halides of boron, aluminium and titanium are not solvolysed by the pure solvent.

Copper acetate is converted into the sulphide, but cupric chloride remains unaffected in liquid hydrogen sulphide. The same is true for acetates of Mn(II), Co(II) and Cd(II), which give the sulphides in contrast to the behaviour of the chlorides[92]. Mercuric acetate gives the sulphide rapidly, although its formation is slow when mercuric chloride is added to liquid hydrogen sulphide.

Amphoteric behaviour has been found for the freshly prepared suspension of arsenic(III) sulphide, which may be dissolved by either acids or bases.

$$As_2S_3 + 3H^+ \rightleftharpoons 2As^{3+} + 3SH^-$$
$$As_2S_3 + 6R_3N + 3H_2S \rightleftharpoons 2[R_3NH]_3[AsS_3]$$

The latter is again decomposed by admission of hydrogen chloride:

$$2[R_3NH]_3[AsS_3] + 6HCl \rightleftharpoons AsS_3 + 3H_2S + 6R_3NHCl$$

Electropositive metals, such as sodium are readily attacked with hydrogen evolution:

$$Na + H_2S \rightarrow Na^+SH^- + \tfrac{1}{2}H_2\uparrow$$

Even noble metals, such as copper, silver and mercury will dissolve in hydrogen sulphide with liberation of hydrogen, showing that due to the poor donor properties of hydrogen sulphide the position of the solvated metal ions is different in liquid hydrogen sulphide from that in water.

Due to the reducing properties of hydrogen sulphide certain redox reactions take place, e.g.,

$$2SeCl_4 + 3H_2S \rightarrow Se_2Cl_2 + 6HCl + 3S \qquad \text{(at } -78°\text{)}$$
$$\text{and} \quad 2SeCl_4 + 4H_2S \rightarrow 2Se + 8HCl + 4S \qquad \text{(at } 15°\text{)}$$
$$FeCl_3 + H_2S \rightarrow 2FeCl_2 + 2HCl + S$$

5. Formamide and Acetamide

Formamide and acetamide are very useful solvents, which are highly associated in the pure liquid states due to hydrogen bonding. Recent reviews have covered their solvent properties in much detail[94,95].

[93] MOLE, R., and R. HOCART: Compt. Rend. Acad. Sci. Paris **230**, 2102 (1950).

[94] DAWSON, L. R.: Part 5, Vol. IV, "Chemistry in Non-Aqueous Ionizing Solvents," Ed. G. JANDER, H. SPANDAU, and C. C. ADDISON, Vieweg and Interscience, Braunschweig, London and New York 1963.

[95] WINKLER, G.: Part 3, Vol. IV, "Chemistry in Non-Aqueous Ionizing Solvents," Ed. G. JANDER, H. SPANDAU, and C. C. ADDISON, Vieweg and Interscience, Braunschweig, London and New York, 1963.

Table 21. *Some Physical Properties of Formamide and Acetamide*

Property	Formamide	Acetamide
Melting Point (°C)	+2.45	+82.0
Boiling Point (°C)	+193.0	+222.0
Dielectric Constant	113.5 (20°)	51.0 (83°)
Spec. Conductivity (Ohm^{-1} · cm^{-1})	6.10^{-7} (20°)	5.10^{-5} (100°)
Viscosity (Centipoise)	0.768 (105°)	1.32 (105°)

Due to their good donor properties formamide and acetamide dissolve many ionic compounds[7,94,96] such as alkali halides[97], various salts of copper, zinc, cadmium, aluminium, tin, lead, nickel, mercury and other acceptor compounds, such as $FeCl_3$ or $SbCl_5$. The high dielectric constants[98] allow considerable dissociation of the compounds ionized in the solutions.

They dissolve also most organic compounds[99] and it is surprising that these solvents are not used to any greater extent. The powerful solvent properties will certainly compensate certain disadvantages, such as the high viscosities, the laborious purification procedure of formamide or the higher temperatures necessary in order to have acetamide available in the liquid state.

Some of the donor-acceptor compounds (solvates) have been known for a long time, such as $(SbCl_5)_2(CH_3CONH_2)_3$[100], the monosolvate of lead chloride[101], the disolvate of cadmium chloride[102] or the solvates of hydrogen halides[7,103,104]. Many other solvates can be prepared[94,95].

In the pure liquid states solvent ions are present due to the autoprotolysis equilibria[105]:

$$2\,HCONH_2 \rightleftharpoons HCONH_3^+ + HCONH^-$$
$$2\,CH_3CONH_2 \rightleftharpoons CH_3CONH_3^+ + CH_3CONH^-$$

It may be noted that the solvated proton has in formamide a lower mobility than the potassium ion, indicating the absence of a special proton transfer mechanism.

The alkali salts of the acid amides behave as bases and may be produced by reacting the alkali amides with the acid amides[106]

$$NaNH_2 + RCONH_2 \rightleftharpoons RCONHNa + NH_3$$

Organic N-bases such as pyridine, α-picoline[9] or triethylamine, also behave as bases in acetamide.

Ammonia is scarcely soluble and behaves as a weak base:

$$NH_3 + RCONH_2 \rightleftharpoons NH_4^+ + RCONH^-$$

[96] WALDEN, P.: Z. Physik. Chem. **46**, 103 (1903).
[97] DAWSON, L. R., and E. J. GRIFFITH: J. Phys. Chem. **56**, 281 (1952).
[98] LEADER, G. R.: J. Amer. Chem. Soc. **73**, 856 (1951).
[99] STAFFORD, O. F.: J. Amer. Chem. Soc. **55**, 3987 (1933).
[100] ROSENHEIM, A., and W. STELLMANN: Ber. dtsch. chem. Ges. **34**, 3377 (1901).
[101] RÖHLER, G.: Z. Elektrochem. **16**, 419 (1910).
[102] PAVLOPOULOS, T., and H. STREHLOW: Z. phys. Chem. (n. F.) **2**, 89 (1954).
[103] VERHOEK, F. H.: J. Amer. Chem. Soc. **58**, 2577 (1936).
[104] JANDER, G., and G. WINKLER: J. Inorg. Nucl. Chem. **9**, 24 (1959).
[105] JANDER, G., and G. WINKLER: J. Inorg. Nucl. Chem. **9**, 32 (1959).
[106] GRÜTTNER, B.: Z. anorg. allg. Chem. **270**, 221 (1952).

Formamide and acetamide are levelling solvents for acids. Solvates of $HClO_4$, HNO_3 and HCl have been isolated. The p_K-values for various organic acids are tabulated in Table 22.

Table 22. p_K-*Values of Organic Acidic Compounds in Formamide*

Picric acid	(1.20)
Trichloroacetic acid	1.46
Dichloroacetic acid	2.85
β-Dibromopropionic acid	4.08
Anilinium ion	4.10
2,6-Dinitrophenol	4.17
o-Nitrobenzoic acid	4.28
Salicylic acid	4.46
Pyridinium ion	4.48
2,4-Dinitrophenol	4.50
Monochloro-acetic acid	4.60
m-Nitrobenzoic acid	5.30
Benzoic acid	6.21
Propionic acid	7.02
Trimethyl acetic acid	7.39
p-Nitrophenol	8.51
Triethylammonium ion	9.99
Piperidinium ion	11.08

Neutralization reactions have been carried out, but there appears to be little information concerning the occurrance of complex reactions.

It is unfortunate that their donor numbers have not been measured, but they should be considered as strong donor solvents and thus may be useful media for the formation of strong complexes.

6. Formic Acid and Acetic Acid

Table 23. *Some Physical Properties of Formic Acid and Acetic Acid*

Property	HCOOH	CH_3COOH
Freezing Point (°C)	+8.4	+16.63
Boiling Point (°C)	+100.8	+118.1
Density at 20° (g · cm^{-3})	1.219	1.049
Mole Volume at 20° (cm^3)	37.82	57.21
Spec. Conductivity at 25° (Ohm^{-1} · cm^{-1})	6.10^{-5}	4.10^{-9}
Dielectric Constant at 20°	57.0	6.1
Dipolemoment (Debye) at 20°	1.19	1.04
Autoprotolysis Constant at 20°	$\sim 10^{-6}$	$\sim 10^{-13}$

Formic acid has better solvent properties than acetic acid, because it has stronger donor properties and a higher dielectric constant. The proton affinity is somewhat smaller than that of acetic acid. Both compounds are associated by hydrogen bonds in the solid state[107]

---

[107] MILLIKAN, R. C., and K. S. PITZER: J. Amer. Chem. Soc. **80**, 3515 (1958).

and probably also to some extent in the pure liquid state. The purification of formic acid[108] is more laborious than that of acetic acid. Water is easily removed from acetic acid by addition of the appropriate amount of acetic anhydride. The chemistry in their solutions has been reviewed recently in considerable detail[109-111].

Both compounds are associated and show autoprotolysis in the pure liquid states:

$$2\,HCOOH \rightleftharpoons [HCOOH_2]^+ + [HCOO]^-$$
$$2\,CH_3COOH \rightleftharpoons [CH_3COOH_2]^+ + [CH_3COO]^-$$

The autoprotolysis constant of formic acid[112,113] is in the order of magnitude of 10^{-6} while that of acetic acid is similar to that of liquid water. The cations formed are called formacidium and acetacidium ions respectively.

Formates and acetates of the alkali metals are easily soluble in the respective solvents and enhance the conductivities of the pure solvents. In formic acid they appear to be appreciably dissociated ($\lambda_0 \sim 80$ at $20°$).

Since the proton affinities of formic and acetic acids are considerably lower than those of water, numerous organic compounds, which show extremely weakly basic properties in water, act as bases in formic and acetic acid since they accept protons more readily than from water; water acts as a base in the acidic solvents:

$$H_2O + HCOOH \rightleftharpoons H_3O^+ + HCOO^-$$

The following p_{K_b}-examples (in parenthesis) show the levelling properties of formic acid for the bases pyridine (2.38), benzidine (2.30), aniline (2.42), glycine (2.38), caffeine (2.30). Urea, triphenylcarbinol, alcohols[114], diethylether, acetanilide and proprionitrile also behave as bases in these media

$$ROH + HCOOH \rightleftharpoons R^+ + H_2O + HCOO^-$$

The low proton affinities are responsible for the small number of proton donors (acids) in these solvents[115]. Hydrogen chloride behaves as a moderately strong

[108] LANGE, J.: Z. phys. Chem. A 187, 27 (1940).

[109] JANDER, G., and G. MAASS: Fortschr. Chem. Forschg. 2, 619 (1953).

[110] HEYMANN, K., and H. KLAUS: Part 1, Vol. IV. "Chemistry in Non-Aqueous Ionizing Solvents," Ed. G. JANDER. H. SPANDAU, and C. C. ADDISON, Vieweg and Interscience, Braunschweig, London and New York 1963.

[111] KNAUER, H.: Part 4, Vol. IV, "Chemistry in Non-Aqueous Ionizing Solvents," Ed. G. JANDER, H. SPANDAU, and C. C. ADDISON, Vieweg and Interscience, Braunschweig, London and New York 1963.

[112] HAMMETT, L. P., and N. DIETZ: J. Amer. Chem. Soc. 52, 4795 (1930).

[113] HAMMETT, L. P., and J. R. DEYRUP: J. Amer. Chem. Soc. 54, 4239 (1932).

[114] ARTHUR, W. R. B., A. G. EVANS, and E. WHITTLE: J. Chem. Soc. 1959, 1940.

[115] JANDER, G., and H. KLAUS: J. Inorg. Nucl. Chem. 1, 334 (1953).

electrolyte both in formic and acetic acids[116]. The strongest acid known is perchloric acid while sulphuric acid and benzene sulphonic acid are considerably
weaker.

$$HClO_4 + HCOOH \rightleftharpoons HCOOH_2^+ + ClO_4^-$$
$$H_2SO_4 + HCOOH \rightleftharpoons HCOOH_2^+ + HSO_4^-$$
$$C_6H_5SO_3H + HCOOH \rightleftharpoons HCOOH_2^+ + C_6H_5SO_3^-$$

Several neutralization-type reactions have been followed by potentiometric and
conductometric techniques, and diethylamine, morphine, urea and triphenylcarbinol have been titrated with toluene sulphonic acid.

Acetic acid has been shown to be a useful solvent for the analytical determination of a number of bases which are too weak in water to be determined by
acidimetric methods. Although the strength of perchloric acid in acetic acid is
smaller than in water, accurate titrations with organic bases are readily carried
out[117-119]. Absolute perchloric acid solutions in glacial acetic acid are easily
prepared by mixing aqueous perchloric acid with the calculated quantity of acetic
anhydride in glacial acetic acid. The solutions can be standardized against sodium
carbonate or sodium phthalate, which give sodium acetate:

$$Na_2CO_3 + CH_3COOH \rightleftharpoons 2 CH_3COO^-Na^+ + H_2O + CO_2$$

The end point may be found potentiometrically or, preferably, by the use of a colour
indicator, such as crystal violet, which gives a sharp colour change from blue to
green. The sharpness of the color change is suppressed by the presence of water,
which therefore should be excluded.

Among the compounds which can be titrated with an accurracy of $\pm 0.2\%$
are aniline, pyridine, methylamine and most other nitrogenous bases[120]. Tertiary
aliphatic amines can be estimated in the presence of primary and secondary
amines[121], since the latter can be converted into neutral amides by addition of
acetic anhydride[122]. It is of particular interest that amino acids can also be determined with great ease in glacial acetic acid[123-125]. Amino sulphamides have been
determined in acetic acid by potentiometric methods[126,127]. Alkali salts of weak
acids, such as picrates, citrates, carbonates or tartrates yield the respective
acetates in acetic acid and thus may be easily titrated with perchloric acid in
this medium[121]. Furthermore derivatives of pyrrole and chlorophyll[128] as well as
polypeptides[129] may be determined in an analogous manner. The reactions of
numerous metal cations have been suggested as having analytical applications[130].

[116] SCHLESINGER, H. I., and A. W. MARTIN: J. Amer. Chem. Soc. 36, 1589 (1914).
[117] HALL, N. F.: Chem. Revs. 8, 191 (1931).
[118] HAMMETT, L. P.: Chem. Revs. 13, 61 (1933).
[119] AUERBACH, G.: Drug Standards 19, 127 (1951).
[120] BLUMRICH, K., and G. BANDEL: Angew. Chem. 54, 373 (1941).
[121] WAGNER, C. D., R. H. BROWN, and F. D. PETERS: J. Amer. Chem. Soc. 69, 2609 (1947).
[122] HASLAM, J., and P. F. HEARN: Analyst 69, 141 (1944).
[123] HARRIS, J. L.: Biochem. J. 29, 2820 (1935).
[124] NADEAU, G. F., and L. E. BRANCHEN: J. Amer. Chem. Soc. 57, 1363 (1935).
[125] TOENNIS, G., and T. P. CALLAN: J. Biol. Chem. 125, 259 (1938).
[126] MARKUNAS, P. C., and J. A. RIDDICK: Analyt. Chem. 23, 337 (1951); 24, 1837 (1952).
[127] TOMIČEK, O.: Coll. Czech. Chem. Comm. 13, 116 (1948).
[128] CONANT, J. B., B. F. CHOW, and E. M. DIETZ: J. Amer. Chem. Soc. 56, 2185 (1934).
[129] HARRIS, L. J.: J. Biol. Chem. 84, 296 (1929).
[130] HARDT, H. D., and M. ECKLE: Z. analyt. Chem. 197, 160 (1963).

Apart from acid-base titrations various addition-, substitution- and redox-reactions have been found to be of analytical interest. Iodine numbers of fats and essential oils may be determined[131], and bromine may be used to titrate organic compounds which can form bromoderivatives[131]. For the titration of phenol with bromine the addition of sodium acetate has been recommended. Redox reagents are in acetic acid chromium(VI) oxide, sodium permanganate, bromine, titanium(III) chloride and chromium(II) salts[132,133]. The titrations are usually carried out in perchloric acid solutions and in an inert atmosphere but traces of water are tolerable.

Formic and acetic acids are very weak donors, probably due to the donor properties at their $C=O$ groups, as is suggested by the existence of solvates with acceptor halides such as $SbCl_3 \cdot CH_3COOH$ [134] and $SnBr_4 \cdot CH_3COOH$ [135]. Infrared spectra of the donor-acceptor-systems $CH_3COOH\text{-}SbCl_5$, $CH_3COOH\text{-}SnCl_4$, $CH_3COOH\text{-}SbCl_3$ and $HCOOH\text{-}SbCl_5$ indicate that the carbonyl oxygen is the donor atom[136] and that the strong acceptor antimony(V) chloride forms adducts with the monomeric acids in which the hydrogen bonds are broken[136]. It was further concluded that hydrogen bonds are retained in the interactions with the weaker acceptors[137] $SnCl_4$ and $SbCl_3$. Boron(III) fluoride forms a 1 : 1 addition compound with acetic acid, which is ionized in the molten state. It has been concluded that the compound is best formulated as $H^+(F_3BOCOCH_3)^{-}$[138].

In the presence of triphenylchloromethane, which is unionized in the pure solvents, chloro-complexes are formed with $SbCl_5$, $FeCl_3$, $SnCl_4$ and other acceptor halides[139].

$$(C_6H_5)_3CCl + MCl_n \rightleftharpoons (C_6H_5)_3C^+MCl_{n+1}$$

The solutions in acetic acid contain scarcely dissociated ion-pairs owing to its low dielectric constant. Some reactions lead to solvolysis products, such as $FeCl(RCOO)_2$. Partial hydrolysis is found to occur with ferric and aluminium chloride, titanium(IV), niobium(V) and tantalum(V)-chlorides, while halides of arsenic(III), zirconium(IV), thorium(IV) and uranium(IV) are completely solvolysed. The high reactivity is undoubtedly due to the presence of acetate ions, and ethylacetate gives many more adducts with acceptor molecules than does acetic acid.

7. Alcohols

Alcohols ROH, which may be considered as derivatives of water, are known to be extremely useful solvents, and only certain aspects of their chemistry can be indicated in this presentation.

[131] Tomiček, O., and J. Dolezal: Acta Pharm. Int. 1, 31 (1950).

[132] Tomiček, O., and J. Heyrovsky: Coll. Czech. Chem. Comm. 15, 997 (1950).

[133] Tomiček, O., and J. Valcha: Coll. Czech. Chem. Comm. 16, 113 (1951).

[134] Ussanovich, M., and T. Ssumarokowa: Zhur. obshch. Khim. USSR 21, 987 (1951); ref. in 110.

[135] Ussanovich, M., W. Klimov, and T. Ssumarokowa: Dokl. Akad. Nauk USSR 113, 364 (1957).

[136] Zackrisson, M., and I. Lindqvist: J. Inorg. Nucl. Chem. 17, 69 (1961).

[137] Kinell, P. O., I. Lindqvist, and M. Zackrisson: Acta Chem. Scand. 13, 1159 (1959).

[138] Greenwood, N. N., R. L. Martin, and H. J. Emeléus: J. Chem. Soc. 1951, 1328.

[139] Cotter, J. L., and A. G. Evans: J. Chem. Soc. 1959, 2988.

Table 24. *Physical Properties of Some Aliphatic Alcohols*

Property	CH_3OH	C_2H_5OH	$n\text{-}C_3H_7OH$	$n\text{-}C_4H_9OH$
Melting Point (°C)	−97.68	−114.5	−126.2	−89.3
Boiling Point (°C)	+64.75	+78.33	+97.4	+117.5
Molecular Volume at 20°	40.72	58.66	74.76	91.08
Dipolemoment (Debye)	1.66	1.70	1.65	1.66
Dielectric Constant at 20°	31.2	25.0	20.7	17.7
Specific Conductivity at 20° ($Ohm^{-1} \cdot cm^{-1}$)	2.10^{-9}	1.10^{-9}	9.10^{-9}	9.10^{-9}

The lower aliphatic alcohols cover reasonable liquid ranges and the dielectric constant is decreased by increasing the chain length of R. The amount of hydrogen bridging is smaller than in water, with which they are miscible.

Autoprotolysis equilibria exist in the pure liquids

$$2\,ROH \rightleftharpoons ROH_2^+ + RO^-$$

with solvated cations and anions; the ionic product of ethyl alcohol has been determined by the EMF method[140] and found to be 10^{-20}.

They dissolve to various degrees ionic compounds such as alkali halides and pseudohalides, nitrates and several 1:2 electrolytes. In the latter group one finds halides of the alkaline earth metals, zinc and cadmium. Since they have donor properties similar to those of water, they readily dissolve many acceptor halides, such as $SbCl_5$, $TiCl_4$, $SnCl_4$, $AlCl_3$ and $AlBr_3$ and solvates may be isolated from the solutions. The solvates are appreciably dissociated in the solutions. Solvates are also known of many other salts, like $MgBr_2$, MgI_2 or $CaCl_2$ which appear to contain solvated cationic species. The compounds $BF_3 \cdot CH_3OH$ and $BF_3(CH_3OH)_2$ have been obtained as colourless crystalline solids. In the molten states they have been formulated as $H^+[F_3BOCH_3]^-$ and $[CH_3OH_2]^+[F_3BOCH_3]^-$ respectively[141].

The solutions are good conductors, but conductivities are greatly increased by the presence of water:

$$CH_3OH_2^+ + H_2O \rightleftharpoons CH_3OH + H_3O^+$$

The ions of the alkali halides show normal mobilities, but the solvated proton ROH_2^+ and the alkoxide ions RO^- have abnormally high mobilities.

When alkali metals are added to an alcohol, an alkali alcoholate is formed which has basic properties.

---

[140] DANNER, P. S.: J. Amer. Chem. Soc. **44**, 2832 (1922).

[141] GREENWOOD, N. N., and R. L. MARTIN: J. Chem. Soc. **1953**, 757.

$$ROH + Na \rightleftharpoons RO^- Na^+ + \tfrac{1}{2} H_2 \uparrow$$

Ammonia and amines also act as bases.

Table 25. *Ionic Mobilities in Methanol and Ethanol*

	CH_3OH at 25°	C_2H_5OH at 25°
ROH_2^+	141.8	57.4
RO^-	53.0	25.4
$[(C_2H_5)_4N]^+$	61.5	27.9
Cl^-	51.2	24.3
ClO_4^-	70.1	33.6
Picrate-ion	46.5	26.3

Table 26. *Basicity Constants in Ethanol*[142] *and Water at 25°*

Base	$-\log K_B$	
	in ethanol	in water
NH_3	8.84	4.77
$C_6H_5NH_2$	13.8	9.3
o-toluidine	13.8	9.5
m-toluidine	13.5	9.8
p-toluidine	13.2	8.7
methyl aniline	14.5	9.7

Most of the acids in aqueous systems also show acidic properties in alcoholic solutions and numerous neutralization reactions can be carried out, which may be studied in a preparative way, by conductometric or potentiometric titrations or by the use of normal colour indicators.

Alcohols are also useful media for the preparation of a number of complex compounds. Thus solutions containing hydrogen chloride may be used for the formation of various chloro-complexes. In general, the chemistry resembles that in water, but solvolysis does not take place to the same extent as it does in water.

---

[142] GOLDSCHMIDT, H.: Z. phys. Chem. **99**, 144 (1921).

Chapter IV

Proton-containing Acceptor Solvents

1. Introduction

The term "acceptor solvent" is as inconsistent as the term "acid." It would be more appropriate to use the terms "acceptor function" and "acidic function," since a molecule may be forced to behave in the opposite way than usual.

"Acceptor solvents" show a tendency to react with electron-pair donors, their donor properties being usually small or non-existent, and although the molecules are polar, no coordination to cations is usually detectable. On the other hand they will be able to accept electron pairs from anions, such as the halide ions, and hence will solvate such anions.

$$SO_2 \; + SO_3^{--} \rightleftharpoons S_2O_5^{--}$$
$$BrF_3 + F^- \quad \rightleftharpoons BrF_4^-$$
$$ICl \; \; + Cl^- \quad \rightleftharpoons ICl_2^-$$

| acceptor solvent | donor anion | solvated anion |

Thus alkali sulphides are soluble in liquid sulphur dioxide, alkali fluorides in bromine(III) fluoride and alkali chlorides in molten iodine monochloride. For many other ionic compounds the energy released in the reaction with an acceptor solvent is often too small to allow reasonable solubilities.

Triphenylcarbonium halides, which are not ionized in donor solvents, will dissolve in acceptor solvents to give solvated halide ions:

$$(C_6H_5)_3CCl + HCl \rightleftharpoons (C_6H_5)_3C^+ + [HCl_2]^-$$
$$(C_6H_5)_3CCl + SO_2 \rightleftharpoons (C_6H_5)_3C^+ + [SO_2Cl]^-$$

Acceptor solvents will furthermore react with amines, phosphines and other donor molecules, e.g.:

$$(C_2H_5)_3N + SO_2 \rightleftharpoons (C_2H_5)_3N : SO_2$$

Acceptor solvents are useful media for the formation of various metal complexes particularly anionic complexes such as halometallates since no competition is provided, by the donor properties of the solvent molecules, for the donor-ligand (which may be called "competitive ligand" in donor solvents).

2. Hydrogen Fluoride

Table 27. *Physical Properties of Hydrogen Fluoride*

Melting Point (°C)	−89.37
Boiling Point (°C)	+19.51
Density at 0° (g · cm⁻³)	1.002
Viscosity at 0° (Centipoise)	0.256
Surface Tension (dynes · cm⁻¹)	10.1
Dielectric Constant at 0°	84.0
Specific Conductivity at 0° (Ohm⁻¹cm⁻¹)	1.10^{-6}
Trouton Constant (cal · mole⁻¹ · deg⁻¹)	24.7
Heat of Vaporization (kcal · mole⁻¹)	7.23

Hydrogen fluoride is a powerful solvent but its use has been severly restricted both by its reactivity towards glass and silica and by its physiological properties. Experimental techniques have, however, been developed which allow the more widespread use of this interesting solvent. Polytetrafluoroethylene and polychlorotrifluoroethylene (the latter being transparent) are now available to handle liquid hydrogen fluoride. Some details have been given in a recent review article[1].

It is hard to understand why hydrogen fluoride is considered an acceptor solvent, because the formation of a coordinate covalent link by accepting an electron pair is not possible. Hydrogen fluoride is nevertheless known to accept fluoride ions to give $[HF_2]^-$ ions by a donor-acceptor reaction

$$\underset{\text{acceptor}}{HF} + \underset{\text{donor}}{F^-} \rightleftharpoons [HF_2]^-$$

The degree of interaction is high enough to consider this a chemical bond, irrespective its actual nature.

Liquid hydrogen fluoride is in fact the strongest hydrogen bonded solvent known, the bond energy of the hydrogen bridge being approximately 8 kcal/mole. Thus the liquid is strongly associated and even in the gas phase hydrogen fluoride is extensively associated; it is probably the most imperfect gas known. Structural arrangements possible are (*a*) zig-zag chains[2, 3] or (*b*) rings[4]; in the solid state the former has been shown to be present[5].

The nature and the extent of molecular association in liquid hydrogen fluoride is influenced by the presence of ionizing impurities, particularly water. Density, acidity and ^1H-n.m.r. behaviour suggest drastic changes in size and arrangement of the $(HF)_n$-polymers.

Liquid hydrogen fluoride is a highly acidic substance, showing a much lower proton affinity in the anhydrous state than in aqueous solution. The ionic species formed by autoprotolysis are also associated and comprise a hydrogen-bonded network. The formal self-ionization is usually represented as

$$2\,HF \rightleftharpoons H_2F^+ + F^-$$

[1] HYMAN, H. H., and J. J. KATZ: Chapter 2 in "Non-Aqueous Solvent Systems," Ed. T. C. WADDINGTON, Academic Press, London-New York 1965.

[2] BRIEGLEB, G., and W. STROHMEIER: Z. Elektrochem. **57**, 668 (1953).

[3] STROHMEIER, W., and G. BRIEGLEB: Z. Elektrochem. **57**, 662 (1953).

[4] FRANCK, E. U., and F. MEYER: Z. Elektrochem. **63**, 571 (1959).

[5] ATOJI, M., and W. N. LIPSCOMB: Acta Cryst. **7**, 173 (1954).

where the actual solvation of the fluoride ion is ignored, and a more appropriate formulation[6] is given as follows:

$$\overset{\displaystyle H^+}{\overbrace{(HF)_n \ + \ (HF)_{m+1}}} \rightleftharpoons [(HF)_nH]^+ + [(HF)_mF]^-$$

$$\underset{\displaystyle F^-}{\underbrace{(HF)_{n+1} + (HF)_m}} \rightleftharpoons [(HF)_nH]^+ + [(HF)_mF]^-$$

This equilibrium can be described as occuring either by autoprotolysis or by transfer of a fluoride ion in the opposite direction[6,7] ("autofluoridolysis"). The liquid structure can be described either by hydrogen bonding or by fluoride bonding between the monomeric units. The existence of the $[HF_2]^-$-ion is well established, but no convincing evidence is available for the existence of compounds containing the $[H_2F]^+$ in the crystalline state.

The autoprotolysis or autofluoridolysis constant has not been measured accurately but has been estimated $K \sim 10^{-12}$ based on the specific conductivity of the purified solvent[8]; the actual value is expected to be still lower, since it is virtually impossible to remove the last traces of water or other conducting impurities.

Both solvent ions show high equivalent conductivities apparently because a chain conduction mechanism is in operation[9,10].

It has been mentioned that anhydrous hydrogen fluoride has strongly acidic properties. While for the reaction in water[11]

$$H_2O + HF \rightleftharpoons H_3O^+ + F^-$$

K_{HF} is 10^{-4}, the proton transfer to an indicator molecule from anhydrous hydrogen fluoride is 10^{17} as great as it is in pure water[1].

Due to its strongly acidic properties hydrogen fluoride acts as a levelling solvent for bases. Metal fluorides with oxidation numbers of less than four behave usually as bases:

$$MF_x + HF \rightleftharpoons [MF_{x-1}]^+ + [HF_2]^-$$

In this process it may be more realistic to assume that fluoride ion transfer is responsible rather than proton transfer. Hydrogen fluoride is indeed a solvent in which an acid is either a proton donor according to BRÖNSTED or a fluoride-ion acceptor, and a base either a proton acceptor or a fluoride ion donor[7].

Alkali metal fluorides, AgF, TlF, SrF_2, BaF_2 and to a lesser extent PbF_2, show reasonable solubilities in anhydrous hydrogen fluoride[12]; they are extensively ionized and, due to the high dielectric constant of anhydrous hydrogen fluoride, practically completely dissociated[9]. Transference measurements indicate that 70% of the current is carried by the fluoride ion.

[6] GUTMANN, V.: Svensk Kem. Tidskr. **68**, 1 (1955).

[7] GUTMANN, V., and I. LINDQVIST: Z. phys. Chem. **203**, 250 (1954).

[8] RUNNER, M. E., G. BALOG, and M. KILPATRICK: J. Amer. Chem. Soc. **78**, 5138 (1956).

[9] KILPATRICK, M., and J. I. LEWIS: J. Amer. Chem. Soc. **78**, 5186 (1956).

[10] HYMAN, H. H., T. I. LANE, and T. A. O'DONNELL: 145th Meeting ACS, Abstracts p. 63 (1963).

[11] ROTH, W. A.: Ann. Chim. Phys. **542**, 35 (1939).

[12] JACHE, A. W., and G. H. CADY: J. Phys. Chem. **56**, 1106 (1952).

Many of the sparingly soluble fluorides are more soluble in the presence of certain complexing agents, such as acetic acid, citric acid, acetonitrile, 1,10-phenanthroline, 8-hydroxyquinoline and carbon monoxide[13].

Ionization of halogen fluorides may again be described by fluoride-ion transfer process[14-16]:

$$ClF_3 \quad + \quad HF \quad \rightleftharpoons \quad [ClF_2]^+ + [HF_2]^-$$
$$BrF_3 \quad + \quad HF \quad \rightleftharpoons \quad [BrF_2]^+ + [HF_2]^-$$
$$\text{base} \qquad \text{acid}$$
$$\text{(F}^-\text{-donor) (F}^-\text{-acceptor)}$$

Hydrogen fluoride acts as an acid because it accepts fluoride ions and the halogen fluorides behave as bases because they donate fluoride ions.

Other covalent fluorides which can act as fairly strong electron pair acceptors, such as antimony(V) fluoride, act as "acids" by accepting fluoride ions from the solvent molecules:

$$SbF_5 \quad + \quad 2\,HF \quad \rightleftharpoons \quad H_2F^+ \quad + \quad SbF_6^-$$
$$\text{acid} \qquad \text{base}$$
$$\text{(F}^-\text{-acceptor) (F}^-\text{-donor)}$$

Various other acceptor fluorides behave in an analogous manner[10,13-17]. Examples are phosphorus(V) fluoride, arsenic(V) fluoride, niobium(V)- and tantalum(V) fluoride, boron(III) fluoride (low solubility), tellurium(IV) fluoride or germanium(IV) fluoride. Thus hydrogen fluoride is an excellent medium for the formation of a number of fluorocomplexes, such as $AgPF_6$, $NaPF_6$, $Ba(PF_6)_2$, $NaAsF_6$, KBF_4, $AgBF_4$, $Ba(TeF_5)_2$, $NaTeF_5$ and many others.

On the other hand titanium(IV) fluoride and silicon(IV) fluoride are nearly insoluble in anhydrous hydrogen fluoride. In the presence of hydrocarbons, titanium(IV) fluoride is made to react and to function as a catalyst in hydrocarbon rearrangement[18-20].

$$R + 2\,TiF_4 + HF \rightarrow RH^+ + [Ti_2F_9]^-$$

The uranium(VI) fluoride-hydrogen fluoride system shows a liquid miscibility gap[21] and no appreciable interaction is indicated in this system[22]. In contrast xenon hexafluoride is easily soluble and undergoes rapid fluorine exchange and extensive dissociation in solution[28].

Neutralization reactions have been followed by conductometric titrations:

$$BrF_2^+HF_2^- + H_2F^+SbF_6^- \rightleftharpoons BrF_2^+ \cdot SbF_6^- + 3\,HF$$

[13] CLIFFORD, A. F., and J. SARGENT: J. Amer. Chem. Soc. **79**, 4041 (1957).

[14] ROGERS, M. T., J. L. SPEIRS, and M. B. PANISH: J. Amer. Chem. Soc. **78**, 3288 (1956).

[15] ROGERS, M. T., J. L. SPEIRS, and M. B. PANISH: J. Phys. Chem. **61**, 366 (1956).

[16] ROGERS, M. T., J. L. SPEIRS, M. B. PANISH, and H. B. THOMPSON: J. Amer. Chem. Soc. **78**, 936 (1956).

[17] KILPATRICK, M., and F. LUBORSKY: J. Amer. Chem. Soc. **76**, 5863 (1954).

[18] McCAULAY, D. A., W. S. HIGLEY, and A. P. LIEN: J. Amer. Chem. Soc. **78**, 3009 (1956).

[19] McCAULAY, D. A., and A. P. LIEN: J. Amer. Chem. Soc. **73**, 2013 (1951).

[20] McCAULAY, D. A., and A. P. LIEN: J. Amer. Chem. Soc. **79**, 2495 (1957).

[21] RUTLEDGE, G. P., R. L. JARRY, and W. DAVIS, Jr.: J. Phys. Chem. **57**, 541 (1953).

[22] MUETTERTIES, E. L., and W. D. PHILLIPS: J. Amer. Chem. Soc. **81**, 1084 (1959).

Due to the poor proton affinity of hydrogen fluoride many substances act as bases because they accept easily protons from the solvent molecules. Water is an obvious example for this type of proton transfer reaction:

$$\overset{\overset{\displaystyle H^+}{\big\lceil\quad\quad\big\rceil}}{H_2O} + 2\,HF \rightleftharpoons H_3O^+ + HF_2{}^-$$

Alcohols, carboxylic acids, aldehydes, ketones, ethers and amines are further examples of basic substances in the liquid hydrogen fluoride solvent system.

Butadiene and other olefinic compounds polymerize in anhydrous liquid hydrogen fluoride or undergo rearrangement reactions. While aliphatic saturated hydrocarbons are usually insoluble, aromatic hydrocarbons are soluble and capable of accepting a proton[23-25]. Acetic acid and even trifluoroacetic acid act as proton acceptors in liquid hydrogen fluoride:

$$\overset{\overset{\displaystyle H^+}{\big\lceil\quad\quad\big\rceil}}{CH_3COOH} + 2\,HF \rightleftharpoons [CH_3COOH_2]^+ + [HF_2]^-$$

Even nitric acid has a higher proton affinity than anhydrous hydrogen fluoride and thus shows basic properties in the latter:

$$\overset{\overset{\displaystyle H^+}{\big\lceil\quad\quad\big\rceil}}{HNO_3} + 2\,HF \rightleftharpoons H_2NO_3{}^+ + HF_2{}^-$$

Dilute solutions of diethyl ether show high electric conductivities, but concentrated solutions are poor conductors. The relative proton affinities remain the same and the extent of proton transfer and hence ionization will also remain unaltered, but the degree of electrolytic dissociation is decreased considerably by increased amounts of diethyl ether because of the lowering of the dielectric constant.

On the other hand perchloric acid behaves as a reasonably strong acid in liquid hydrogen fluoride.

$$\overset{\overset{\displaystyle H^+}{\big\lceil\quad\quad\big\rceil}}{HClO_4} + HF \rightleftharpoons ClO_4{}^- + H_2F^+$$

It may be noted that hydrogen chloride, hydrogen bromide and hydrogen iodide are nearly insoluble in liquid hydrogen fluoride[26] but their solubilities are sufficient to precipitate silver halides from solutions of silver fluoride[27]. Xenon(IV) fluoride is sparingly soluble in anhydrons hydrogen fluoride but xenon(VI) fluoride is very soluble and extensively ionized[28].

Nitrobenzene functions as an acid-base indicator both in sulphuric acid and in hydrofluoric acid.

[23] KILPATRICK, M., and F. LUBORSKY: J. Amer. Chem. Soc. **75**, 577 (1953).
[24] MACKOR, E. L., A. HOFSTRA, and J. H. VAN DE WAALS: Trans. Farad. Soc. **54**, 186 (1958).
[25] MacLEAN, C., and E. L. MACKOR: Disc. Farad. Soc. **34**, 165 (1962).
[26] FREDENHAGEN, K., and G. CADENBACH: Z. phys. Chem. **A 146**, 245 (1930).
[27] FREDENHAGEN, K.: Z. anorg. allg. Chem. **242**, 23 (1939).
[28] HYMAN, H. H., and L. A. QUARTERMAN, in: "Noble Gas Compounds," Ed. H. H. HYMAN, p. 275, Univ. of Chicago Press 1963.

Despite the obvious reactivity of hydrogen fluoride a number of carbohydrates and proteins are easily soluble and can be frequently recovered unchanged from the solutions. It is important that the heat evolved in the course of dissolutions is dissipated. Cellulose[29] forms conducting solutions and the material recovered is easily hydrolysed to give glucose.

Proteins including fibrous proteins, are normally insoluble in water but can usually be dissolved in anhydrous hydrogen fluoride without difficulty. Examples are: ribonuclease, insuline, trypsine, serum albumine, serum globuline, haemoglobine and collagene. Insuline retains essentially its full biological activity when recovered from the solution[29]. The same is true for the enzymes ribonuclease and lysozyme provided they are exposed to liquid hydrogen fluoride only for short periods at low temperature[30].

Hydrogen fluoride swells collagen, but probably does not break the peptide bonds. N-carboxy anhydrides are polymerized in hydrogen fluoride to give poly-α-amino acids[31] with chain length of about 30 units.

Rearrangement reactions occur with serine- and threonine-containing peptides.

Chlorophyll is soluble and so is vitamin B_{12}, the latter giving a deep olive green solution in hydrogen fluoride. The Co(III) coordination compound is recovered from the solutions, from which vitamin B_{12} can be regenerated with its full B_{12} activity.

3. Liquid Hydrogen Chloride, Hydrogen Bromide and Hydrogen Iodide

Table 28. *Physical Properties of Hydrogen Chloride, Hydrogen Bromide and Hydrogen Iodide*

Property	HCl	HBr	HI
Melting Point (°C)	−114.6	−88.5	−30.9
Boiling Point (°C)	−84.1	−67.0	−35.0
Heat of Vaporization (kcal · mole^{-1}) ...	3.86	4.21	4.72
Trouton Constant (cal · mole^{-1} · deg^{-1}) .	20.4	20.4	19.8
Dielectric Constant	9.3	7.0	3.4
Spec. Conductivity (Ohm^{-1}cm^{-1})	3.10^{-8}	1.10^{-8}	8.10^{-9}
	(−80°)	(−84°)	(−45°)

Hydrogen chloride, hydrogen bromide and hydrogen iodide have low and limited liquid ranges[32]. The low boiling points and the normal Trouton constants show that in the liquid states they are not appreciably associated and thus not hydrogen bonded to any measurable extent. Their dielectric constants are low and so are the specific conductivities of the pure liquids. They are typical acceptor solvents showing small tendencies to solvate cations and indeed most of the metal salts are nearly insoluble in the pure liquid hydrogen halides. On the other hand they tend to solvate anions and to react with various electron pair donors. The compounds $(C_2H_5)_2O \cdot HCl$, $(C_2H_5)_2O \cdot (HCl)_2$, $(C_2H_5)_2O(HCl)_5$, $CH_3CN \cdot HCl$, $(CH_3CN)_2(HCl)_3$,

[29] KATZ, J. J.: Archs. Biochem. Biophys. **61**, 293 (1954).

[30] KOCH, A. L., W. A. LAMONT, and J. J. KATZ: Archs. Biochem. Biophys. **63**, 106 (1956).

[31] KOPPLE, K. D., and J. J. KATZ: J. Amer. Chem. Soc. **78**, 6199 (1956).

[32] PEACH, M. E., and T. C. WADDINGTON: Chapter 3 in "Non-Aqueous Solvent Systems," Ed. T. C. WADDINGTON, Acad. Press, New York-London 1965.

$CH_3CN(HCl)_5$ and $CH_3CN(HCl)_7$ have been described. Since water is a stronger proton acceptor than the hydrogen halides, it forms $H_3O^+X^-$, which are insoluble in the liquid hydrogen halides as are most other ionic compounds. The presence of self-ionization equilibria of the following type has been postulated[32],

$$3\,HX \rightleftharpoons H_2X^+ + HX_2^-$$

by analogy to the behaviour of liquid hydrogen fluoride, but conclusive evidence for the existence of the solvent cations H_2X^+ has not been produced.

On the other hand the existence of the solvent anions, for example $[HCl_2]^-$ is well established, since it is found in crystalline tetramethylammonium hydrogendichloride[33, 34] and cesium hydrogen dichloride[35], and IR-frequencies have been assigned to the anion[34, 36] which is believed to be linear[36, 37]. Both the $[HCl_2]^-$ and $[HBr_2]^-$-ions have been obtained in the tropylium compounds[38] $[C_7H_7]^+[HX_2]^-$ and the compounds $[(C_6H_5)_4As][HBr_2]$[39] and $[(C_4H_9)_4N][HI_2]$[40] have also been characterized. The formation of such ions may be regarded as due to donor-acceptor reactions:

$$\underset{\substack{\text{acceptor-}\\\text{solvent}}}{HX} + \underset{\text{donor}}{X^-} \rightleftharpoons [HX_2]^-$$

In principle two different definitions can be given for acids and bases analogous to those used in the liquid hydrogen fluoride solvent system. An acid may be described as either a proton donor or a halide ion acceptor and a base as a proton acceptor or a halide ion donor[6, 7].

Due to the high acidities of the solvents it is difficult to find protonic acids in these systems. Covalent chlorides, such as boron(III) chloride, behave as acids in liquid hydrogen chloride by accepting a chloride ion and they can be neutralized by tetraalkylammoniumhydrogendichloride:

$$[(CH_3)_4N]^+[HCl_2]^- + BCl_3 \rightleftharpoons [(CH_3)_4N]^+[BCl_4]^- + HCl$$
$$[(CH_3)_4N]^+[HCl_2]^- + BF_3 \rightleftharpoons [(CH_3)_4N]^+[BF_3Cl^-] + HCl$$

The occurrance of such reactions has been established by conductometric measurements.

Phosphorus(V) chloride was found to behave as a base in hydrogen chloride as well as in the proton-free acceptor solvents iodine(I) chloride or arsenic(III) chloride:

$$Cl^-$$

$$PCl_5 + HCl \rightleftharpoons [PCl_4]^+ + [HCl_2]^-$$
$$PCl_5 + ICl \rightleftharpoons [PCl_4]^+ + [ICl_2]^-$$
$$PCl_5 + AsCl_3 \rightleftharpoons [PCl_4]^+ + [AsCl_4]^-$$

$$\underset{\text{donor}}{\quad} \quad \underset{\substack{\text{acceptor-}\\\text{solvent}}}{\quad}$$

[33] WADDINGTON, T. C.: Trans. Farad. Soc. **54**, 25 (1958).

[34] WADDINGTON, T. C.: J. Chem. Soc. **1958**, 1708.

[35] VALLEÉ, R. E., and D. H. McDANIEL: J. Amer. Chem. Soc. **84**, 3412 (1962).

[36] SHARP, D. W. A.: J. Chem. Soc. **1958**, 2558.

[37] CHANG, S., and E. F. WESTRUM, Jr.: J. Chem. Phys. **36**, 2571 (1962).

[38] HARMON, K. M., and S. DAVIS: J. Amer. Chem. Soc. **84**, 4359 (1962).

[39] WADDINGTON, T. C., and J. A. WHITE: Proc. Chem. Soc. **1960**, 85.

[40] SALTHOUSE, J. A., and T. C. WADDINGTON: J. Chem. Soc. **1966**, (A), 1188.

Most of the conclusions have been drawn from measurements of the electrical conductivities. This, however, although providing some indication of the number and mobility of the ions present, gives little information as to their nature.

The low proton affinities of the solvent molecules allows the development of basic properties for a number of proton acceptors[39, 41, 42]. Carbon-halogen compounds, in which the carbon-halogen bond is easily broken with the formation of a stable carbonium ion, act as bases[42, 43] to give ionized solutions of high conductivities:

$$
\begin{array}{ll}
& \overset{\displaystyle Cl^-}{\overbrace{}\!\!\downarrow} \\
(C_6H_5)_3CCl & + \quad HCl \rightleftharpoons [(C_6H_5)_3C]^+ + [HCl_2]^- \\
(C_6H_5)_3CCl & + \quad HI \;\; \rightleftharpoons [(C_6H_5)_3C]^+ + [HClI]^- \\
\text{Cl}^- \text{ donor} & \text{acceptor-} \\
& \text{solvent}
\end{array}
$$

Triphenylamine is a strong base in liquid hydrogen chloride[43, 44] and hydrazobenzene is protonated with subsequent rearrangement to give benzidine[43]. Triphenylphosphine is protonated in all three liquid hydrogen halides[42, 43] and can be reacted with boron halides

$$(C_6H_5)_3P + HX + BX_3 \rightleftharpoons (C_6H_5)_3PH^+BX_4^-$$

but the analogous reaction with triphenylarsine leads to $(C_6H_5)_3As \cdot BCl_3$. In general hydrogen halides are levelling solvents for bases.

As might be expected from the strong acidic properties of the solvents protonation occurs with alcohols, ethers[45, 46], nitrocompounds, sulphones, alkyl sulphides, nitriles as well as with π-bonded systems[32]:

$$
\begin{aligned}
CH_3C &\equiv N + 2\,HCl \rightarrow CH_3C \equiv NH^+ + HCl_2^- \\
(C_6H_5)_2C &= CH_2 + 2\,HCl \rightarrow (C_6H_5)_2CCH_3^+ + HCl_2^- \\
C_6H_5C &\equiv CH + 4\,HCl \rightarrow C_6H_5CH_3C^{2+} + 2\,HCl_2^-
\end{aligned}
$$

Phosphorus(V) fluoride reacts in liquid hydrogen chloride with chloride ion donors to give hexafluorophosphate:

$$2\,(CH_3)_4NCl + 3\,PF_5 \rightleftharpoons 2\,(CH_3)_4NPF_6 + PF_3Cl_2$$

When phosphorus(V) chloride is reacted with phosphorus(V) fluoride in the same solvent both ionic $PCl_4^+PF_6^-$ and covalent PF_3Cl_2 are formed, which have the same chemical composition:

$$2\,PCl_5 + 3\,PF_5 \rightarrow 2\,PCl_4^+PF_6^- + PF_3Cl_2$$

Arsenic(V) fluoride is solvolysed in liquid hydrogen chloride to give tetrachloroarsonium hexafluoroarsenate:

$$2\,AsF_5 + 4\,HCl \rightleftharpoons AsCl_4^+AsF_6^- + 4\,HF$$

[41] WADDINGTON, T. C., and J. A. WHITE: J. Chem. Soc. **1963**, 2701.
[42] KLANBERG, F., and H. W. KOHLSCHÜTTER: Z. Naturf. **16 b**, 69 (1961).
[43] PEACH, M., E., and T. C. WADDINGTON: J. Chem. Soc. **1961**, 1238.
[44] PEACH, M., E., and T. C. WADDINGTON: J. Chem. Soc. **1962**, 600.
[45] MAAS, O., and D. McINTOSH: J. Amer. Chem. Soc. **35**, 535 (1913).
[46] McINTOSH, D.: J. Amer. Chem. Soc. **30**, 1097 (1908).

It may be noted that the free halogens will not react with the respective liquid hydrogen halides. Not even iodine reacts with hydrogen iodide, where the formation of a polyiodide anion might be expected. On the other hand iodine monochloride will react with chloride ion donors in hydrogen chloride just as well as in the absence of a solvent[32]

$$R_4N^+Cl^- + ICl \rightleftharpoons R_4N^+ICl_2^-$$

There are some indications that hydrogen bromide and hydrogen iodide can act as acids in liquid hydrogen chloride.

$$\overset{\displaystyle Cl^-}{\overset{\displaystyle \lceil\quad\quad\downarrow}{HCl_2^- + H\overset{}{B}r}} \rightleftharpoons HCl + HClBr^-$$

Triphenylcarbinol is solvolysed to the ionized species $(C_6H_5)_3C^+HCl_2^-$ [43]:

$$(C_6H_5)_3COH + 3HCl \rightleftharpoons (C_6H_5)_3C^+HCl_2^- + H_3O^+Cl^-$$

Hydroxonium chloride is nearly insoluble, just as water is scarcely soluble in liquid hydrogen chloride.

Triphenylstannyl chloride undergoes complete solvolysis[32]:

$$(C_6H_5)_3SnCl + HCl \rightarrow (C_6H_5)_2SnCl_2 + C_6H_6$$

by an unknown mechanism.

Other solvolytic reactions are:

$$SbF_3 + 3HCl \rightleftharpoons SbCl_3 + 3HF$$
$$BCl_3 + 3HI \rightleftharpoons BI_3 + 3HCl$$

Many of the reactions which have been mentioned are not specific for the solvent-systems under consideration and are expected also to take place in other acceptor solvents.

4. Liquid Hydrogen Cyanide

Liquid hydrogen cyanide has a limited liquid range, in which the degree of association is apparently similar to that of the hydrogen halides HX, where X = Cl, Br, I. On the other hand it has in common with liquid hydrogen fluoride a high dielectric constant and it has been used in the study of conductivities in non-aqueous solutions[47].

Table 29. *Some Physical Properties of Hydrogen Cyanide*

Melting Point (°C)	−13.35
Boiling Point (°C)	+25.0
Density at 25° (g · cm^{-3})	0.716
Dielectric Constant at 0°	153.0
Dielectric Constant at 25°	123.0
Specific Conductivity at 0° (Ohm^{-1}cm^{-1})	8.10^{-9}
Heat of Vaporization (kcal · mole^{-1})	6.74

[47] LANGE, J., J. BERGÅ, and N. KONOPIK: Mh. Chem. **80**, 708 (1949).

It is a hazardous solvent to use due to its toxicity combined with a high volatility.

The small conductivity of the purified liquid[47] is regarded in part to be due to autoprotolysis,

$$2\,HCN \rightleftharpoons H_2CN^+ + CN^-$$

but no compound has been prepared containing the H_2CN^+ ion in the crystalline state. The existence of the autoprotolysis equilibrium is in agreement with the results of semiquantitative potentiometric measurements[48], which suggest an autoprotolysis constant in the order of magnitude of 10^{-18}.

Hydrogen cyanide is a poor solvent because its coordinating properties both as a donor and as an acceptor are low, although it has a high dielectric constant. There is also no conclusive evidence for the solvation of the cyanide anion, in contrast to that available even in hydrogen bromide.

Cyanides are solvo-bases in the liquid hydrogen cyanide solvent system. Potassium cyanide is moderately soluble and is dissociated due to the high dielectric constant of the solvent. Because this solvent lacks any extensive hydrogen bonded structure and because of its low acceptor properties little energy is provided by solvation; ionic compounds having high lattice energies are, accordingly, not appreciably soluble and, apart from potassium cyanide, alkali metal cyanides are practically insoluble. Solutions of cyanides are unstable since they promote the occurrance of polymerization reactions. Due to the proton-donor properties of the solvent organic nitrogen bases and other organic compounds will accept a proton from the solvent and therefore act as bases. Examples are amylamine and strychnine, which give highly conducting solutions, while pyridine gives solutions of very low conductivities ($\lambda_0 \sim 10^{-1}$). Soluble iodides give highly conducting solutions[48] such as KI ($\lambda_c \sim 300$), Me$_4$NI ($\lambda_c \sim 320$) both measured at $c \sim 0.1$ M.

Strong acids in water, like sulphuric acid, hydrogen chloride, or nitric acid, behave as very weak acids in liquid hydrogen cyanide. Colour indicators may be used in freshly prepared solutions[49] but dichloroacetic acid will not cause the indicator molecules to give the colour characteristic for acidic solutions.

Neutralization reactions have been followed by various techniques such as by conductometric and potentiometric methods[49, 50]. It is, however, difficult to estimate the extent to which the solvent is actually taking part in the reactions and to what extent it is acting merely as a diluent.

Table 30. *Neutralization Reactions in Liquid Hydrogen Cyanide*

Base	Acid	observed Reaction Products
KCN	H_2SO_4	HSO_4, K_2SO_4 (insoluble)
$(C_2H_5)_3N$	H_2SO_4	$(C_2H_5)_3NH^+HSO_4^-$, $(C_2H_5)_3NH_2{}^{2+}+SO_4{}^{2-}$
$(C_2H_5)_3N$	HCl	$(C_2H_5)_3NH^+Cl^-$
C_5H_5N	HCl	$C_5H_6N^+Cl^-$
C_5H_5N	H_2SO_4	$C_5H_6N^+HSO_4^-$, $C_5H_7N_2{}^+SO_4{}^{2-}$
$(C_2H_5)_3N$	Picric acid	$[(C_2H_5)_3NH][picrate]$
$(C_2H_5)_3N$	CCl_2HCOOH	$[(C_2H_5)_3NH][CCl_2HCOO]$
KCN	HNO_3	KNO_3

[48] JANDER, G., and G. SCHOLZ: Z. phys. Chem. **A 192**, 163 (1943).
[49] JANDER, G., and B. GRÜTTNER: Chem. Ber. **81**, 102 (1948).
[50] JANDER, G., and B. GRÜTTNER: Chem. Ber. **81**, 114 (1948).

Silver salts are converted into insoluble silver cyanide:

$$AgNO_3 + HCN \rightarrow AgCN \downarrow + HNO_3$$
$$Ag_2SO_4 + 2\,HCN \rightarrow 2\,AgCN \downarrow + H_2SO_4$$
$$Ag_3PO_4 + 3\,HCN \rightarrow 3\,AgCN \downarrow + H_3PO_4$$

but freshly precipitated silver cyanide dissolves in the presence of bases[50]:

$$AgCN + (C_2H_5)_3N + HCN \rightarrow [(C_2H_5)_3NH][Ag(CN)_2]$$
$$AgCN + KCN \rightarrow K[Ag(CN)_2]$$

Acid halides are solvolysed in hydrogen cyanide:

$$RCOCl + HCN \rightleftharpoons RCOCN + HCl$$

Ferric chloride gives complex cyanides in the presence of bases:

$$FeCl_3 + 6\,(C_2H_5)_3N + 6\,HCN \rightleftharpoons [(C_2H_5)_3H]_3[Fe(CN)_6] + 3\,(C_2H_5)_3NHCl$$

Analogous behaviour is found for freshly precipitated mercuric perchlorate:

$$Hg(ClO_4)_2 + KCN + 2\,HCN \rightleftharpoons K[Hg(CN)_3] + 2\,HClO_4$$

5. Sulphuric Acid

Table 31. *Some Physical Properties of Anhydrous Sulphuric Acid*

Melting Point (°C)	+10.371
Boiling Point (°C)	290 − 317
Density at 25° (g · cm^{-3})	1.8269
Viscosity (Centipoise) at 25°	24.54
Heat of Fusion (kcal · mole^{-1})	2.5
Dielectric Constant at 10°	120.0
Specific Conductance at 25° (Ohm^{-1} · cm^{-1}) .	1.044 · 10^{-2}

Although excellent review articles are available[51-55], some characteristic features of coordination chemistry in this solvent will be described briefly.

Sulphuric acid is a viscous liquid which is associated in the liquid state by means of strong hydrogen bonds. It is a good solvent to ionize and, subsequently, to dissociate many compounds due to its coordinating properties and high dielectric constant.

The acceptor properties are reflected in its poor ability to solvate metal cations and in its capacity to allow complete dissociation of triphenylcarbonium salts; the presence of SO_3 produced by self dehydration (see below) may be responsible —at least in part—for the acceptor properties.

[51] GILLESPIE, R. J., and E. A. ROBINSON: Chapter 4 in "Non-Aqueous Solvent Systems," Ed. T. C. WADDINGTON, Academic Press, New York-London 1965.

[52] GILLESPIE, R. J., and J. A. LEISTEN: Quart. Revs. **8**, 40 (1954).

[53] GILLESPIE, R. J.: Rev. Pure and Appl. Chem. (Australia) **9**, 1 (1959).

[54] GILLESPIE, R. J., and E. A. ROBINSON: Advances of Inorganic Chemistry and Radiochemistry, Ed. H. J. EMELÉUS, and A. G. SHARPE, **1**, 385 (1959); Acad. Press, New York.

[55] SISLER, H. H.: "Chemistry in Non-Aqueous Solvents," Reinhold Publ. Corp., New York-London 1964.

The solid has a layer-type structure[56], in which each sulphuric acid molecule is hydrogen bonded to four others. It has been assumed that this association persists to a great extent in the liquid state just as the structure of liquid water has been related to that of ice, although recent n.m.r. work in water suggests drastic structural differences between water and ice[57].

The high degree of ionization in the pure liquid is indicated by its high conductivity and by the freezing point diagram, which shows a flat maximum at $+10.37°$, while the extrapolated figure for the melting point of the hypothetically completely undissociated acid is $+10.625$ [58]. Sulphuric acid is subject to autoprotolysis to an appreciable extent according to[59]:

$$H_2SO_4 + H_2SO_4 \rightleftharpoons H_3SO_4^+ + HSO_4^-; \quad K_1 \sim 2 \cdot 7.10^{-4}$$

In addition to autoprotolysis a self-dehydration equilibrium is operative in liquid sulphuric acid.

$$2\,H_2SO_4 \rightleftharpoons H_2S_2O_7 + H_2O$$

Since water is appreciably more basic than sulphuric acid, it is ionized according to:

$$H_2O + H_2SO_4 \rightleftharpoons H_3O^+ + HSO_4^-$$

while disulphuric acid acts as a proton donor towards sulphuric acid:

$$H_2S_2O_7 + H_2SO_4 \rightleftharpoons HS_2O_7^- + H_3SO_4^+$$

Since HSO_4^- and $H_3SO_4^+$ ions are in equilibrium as a result of the autoprotolysis, it follows that H_3O^+ and $HS_2O_7^-$ must also be in equilibrium[60]:

$$2\,H_2SO_4 \rightleftharpoons H_3O^+ + HS_2O_7^-; \quad K_2 = 5.10^{-5} \text{ at } 25°$$

which is called the ionic self-dehydration reaction.

The autodeuterolysis constant of D_2SO_4 is smaller than the autoprotolysis constant of H_2SO_4 showing that D_2SO_4 is a weaker acid than H_2SO_4. Similarly $D_2S_2O_7$ is a weaker acid in D_2SO_4 than is $H_2S_2O_7$ in H_2SO_4 [51].

Solutions of hydrogen sulphates show high conductivities[61-64] which are, at high dilution, practically identical[63, 65]. This suggests that the HSO_4^- ion carries most of the current and transfer number measurements have confirmed this

[56] PASCARD, R.: Compt. Rend. Acad. Sci. Paris **108**, 249 (1956).
[57] KITT, J. W., and T. H. LILLEY: Chem. Comm. **1967**, 323.
[58] BASS, S. J., and R. J. GILLESPIE: J. Chem. Soc. **1960**, 814.
[59] GILLESPIE, R. J.: J. Chem. Soc. **1950**, 2493.
[60] GILLESPIE, R. J.: J. Chem. Soc. **1950**, 2516.
[61] GILLESPIE, R. J., J. V. OUBRIDGE, and C. SOLOMONS: J. Chem. Soc. **1956**, 1804.
[62] HAMMETT, L. P., and F. A. LOWENHEIM: J. Amer. Chem. Soc. **56**, 2620 (1934).
[63] BASS, S. J., R. J. GILLESPIE, and J. V. OUBRIDGE: J. Chem. Soc. **1960**, 837.
[64] BASS, S. J., R. J. GILLESPIE, and E. A. ROBINSON: J. Chem. Soc. **1960**, 821.
[65] GILLESPIE, R. J., and S. WASIF: J. Chem. Soc. **1953**, 209.

conclusion[65-67]. Only a few percent of the current is due to the cation. The $H_3SO_4^+$ shows an abnormally high mobility[68, 69].

Table 32. *Apparent Transport Numbers of Some Cations in H_2SO_4 at 25°*

Ion	c	transport number
Ag^+	0.249	0.026
K^+	0.62	0.030
Na^+	0.79	0.021
Li^+	0.56	0.013
Ba^{2+}	0.17	0.009
Sr^{2+}	0.21	0.003

Although ionic mobilities are small due to the high viscosity of the solvent medium, the mobilities of $H_3SO_4^+$ and HSO_4^- are appreciably higher than those of most ions in water. The high viscosity of the solvent does not in fact affect the mobilities of such ions, because a proton transfer mechanism is in operation analogous to that accepted for H_3O^+ and OH^- in water.

Most compounds which are soluble in sulphuric acid behave as electrolytes because they are sufficiently basic to form strong hydrogen bonds with sulphuric acid and it is likely that proton transfer along the hydrogen bonds will occur, resulting in ionization. The high dielectric constant of the solvent favours fairly complete dissociation of the ionized compounds. Bases give the hydrogen sulphate ion HSO_4^- and due to the low proton affinity of sulphuric acid, bases are the largest class of electrolytes. Many substances indeed act as bases in sulphuric acid, although they may not show such properties in water. Sulphuric acid has a levelling effect on the strength of bases, just as liquid ammonia has on the strength of acids.

Thus sulphuric acid is a valuable solvent in the study of the protonation of very weak bases such as ketones and nitro compounds and in the preparation of stable solutions of reactive ions, such as carbonium ions, NO_2^+, and I_3^+, which are unstable in water.

Examples of bases in sulphuric acid are alkali sulphates and hydrogen sulphates

$$KHSO_4 \rightleftharpoons K^+ + HSO_4^-$$
$$K_2SO_4 + H_2SO_4 \rightleftharpoons 2K^+ + 2HSO_4^-$$

or compounds which are easily solvolysed:

$$NH_4ClO_4 + H_2SO_4 \rightleftharpoons NH_4^+ + HSO_4^- + HClO_4$$
$$Na_3PO_4 + 3H_2SO_4 \rightleftharpoons 3Na^+ + 3HSO_4^- + H_3PO_4$$
$$KNO_3 + H_2SO_4 \rightleftharpoons K^+ + HSO_4^- + HNO_3$$

Another group includes ketones, carboxylic acids, esters, amines, amides, phosphines, nitro compounds, nitriles and others.

[66] GILLESPIE, R. J., and S. WASIF: J. Chem. Soc. **1953**, 215.
[67] GILLESPIE, R. J., and S. WASIF: J. Chem. Soc. **1953**, 221.
[68] FLOWERS, R. H., R. J. GILLESPIE, and E. A. ROBINSON: Canad. J. Chem. **58**, 1363 (1960).
[69] HALL, S. K., and E. A. ROBINSON: Canad. J. Chem. **42**, 1113 (1964).

$$H^+$$

$$R_2CO + H_2SO_4 \rightleftharpoons R_2COH^+ + HSO_4^-$$
$$RCOOH + H_2SO_4 \rightleftharpoons RCOOH_2^+ + HSO_4^-$$
$$RCOOR' + H_2SO_4 \rightleftharpoons RCOOR'H^+ + HSO_4^-$$
$$RNH_2 + H_2SO_4 \rightleftharpoons RNH_3^+ + HSO_4^-$$
$$R_3P + H_2SO_4 \rightleftharpoons R_3PH^+ + HSO_4^-$$
$$RNO_2 + H_2SO_4 \rightleftharpoons RNO_2H^+ + HSO_4^-$$

Nitric acid behaves as a base according to the reactions:

$$HNO_3 + H_2SO_4 \rightleftharpoons NO_2^+ + HSO_4^- + H_2O$$
$$H_2O + H_2SO_4 \rightleftharpoons H_3O^+ + HSO_4^-$$

Phosphoric acid and selenium dioxide act also as bases,

$$H_3PO_4 + H_2SO_4 \rightleftharpoons H_4PO_4^+ + HSO_4^-$$
$$SeO_2 + H_2SO_4 \rightleftharpoons HSeO_2^+ + HSO_4^-$$

as well as triphenylcarbinol and mesitoic acid. Hydroxocompounds and alcohols are dehydrated and behave also as bases:

$$XOH + 2H_2SO_4 \rightleftharpoons XSO_4H + H_3O^+ + HSO_4^-$$
$$ROH + 2H_2SO_4 \rightleftharpoons RSO_4H + H_3O^+ + HSO_4^-$$

There are only a few known substances which act as acids in sulphuric acid. Even perchloric acid, the strongest mineral acid in water, is undissociated in sulphuric acid despite the high dielectric constant. Disulphuric acid and higher polysulphuric acids are definitely acidic as well as the complex acid $HB(HSO_4)_4$, which is obtained by dissolving boric acid in oleum[70].

$$H_3BO_3 + 3H_2S_2O_7 = H_3SO_4^+ + B(HSO_4)_4^- + H_2SO_4$$

A mixture of HNO_3 and H_2SO_4 is used for nitration reactions since the nitronium ion is produced by dehydration of nitric acid in the presence of sulphuric acid:

$$HNO_3 + 2H_2SO_4 \rightleftharpoons NO_2^+ + 2HSO_4^- + H_3O^+$$

N_2O_5 is dissociated in sulphuric acid and the following equations describe a possible subsequent reaction sequence:

$$N_2O_5 \rightleftharpoons NO_2^+ + NO_3^-$$
$$NO_3^- + H_2SO_4 \rightleftharpoons HNO_3 + HSO_4^-$$
$$HNO_3 + H_2SO_4 \rightleftharpoons NO_2^+ + HSO_4^- + H_2O$$
$$H_2O + H_2SO_4 \rightleftharpoons H_3O^+ + HSO_4^-$$

$$\overline{N_2O_5 + 3H_2SO_4 \rightleftharpoons 2NO_2^+ + 3HSO_4^- + H_3O^+}$$

In an analogous manner N_2O_4 and N_2O_3 are dissociated in sulphuric acid:

$$N_2O_4 + 3H_2SO_4 \rightleftharpoons NO_2^+ + NO^+ + H_3O^+ + 3HSO_4^-$$
$$N_2O_3 + 3H_2SO_4 \rightleftharpoons 2NO^+ + H_3O^+ + 3HSO_4^-$$

[70] THOMPSON, A., and N. N. GREENWOOD: J. Chem. Soc. **1959**, 736.

Due to the high mobilities of the solvent ions acid-base reactions can conveniently be followed by conductometric titrations, for example:

$$KHSO_4 + HB(HSO_4)_4 \rightleftharpoons KB(HSO_4)_4 + H_2SO_4$$

Here, the conductivity minimum occurs at a molar ratio of 0.98 instead of 1.0, but attempts to isolate the potassium salt were not successful[73].

Since only a few very weak acids are available in the sulphuric acid solvent system, solvolysis reactions are the common type of acid-base reactions:

$$NO_2ClO_4 + H_2SO_4 \rightleftharpoons HClO_4 + NO_2^+HSO_4^-$$

It has been claimed that, by the reaction of boron(III) chloride, $HB(HSO_4)_4$ is obtained as a "wet solid"[70],

$$BCl_3 + 4H_2SO_4 = HB(HSO_4)_4 + 3HCl$$

but examination of the Raman spectra also indicates the presence of B-O-B links[71]. Salts of sulphatoboric acid may, however, be prepared[72] according to either

$$(NH_4)_2SO_4 + B(OH)_3 + 6SO_3 = 2NH_4[B(SO_4)_2] + 3H_2SO_4$$

or $\quad BN + 2H_2SO_4 = NH_4[B(SO_4)_2].$

Complex sulphato-compounds of boron and silicon are obtained due to an instability of $[B(HSO_4)_4]^-$ analogous to that of $[B(OH)_4]^-$[73]:

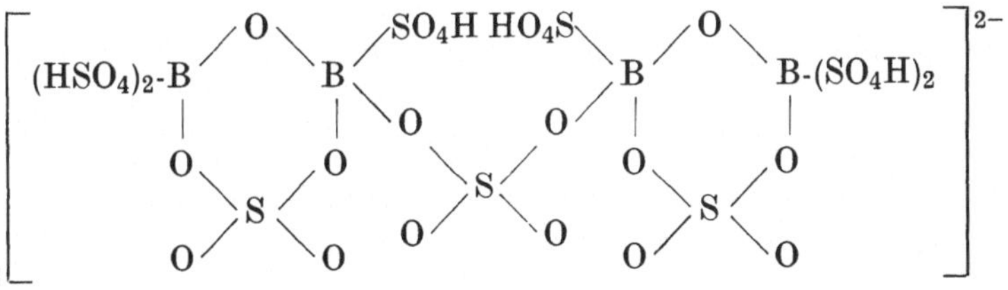

Polymers of silicon have been formulated as[74]:

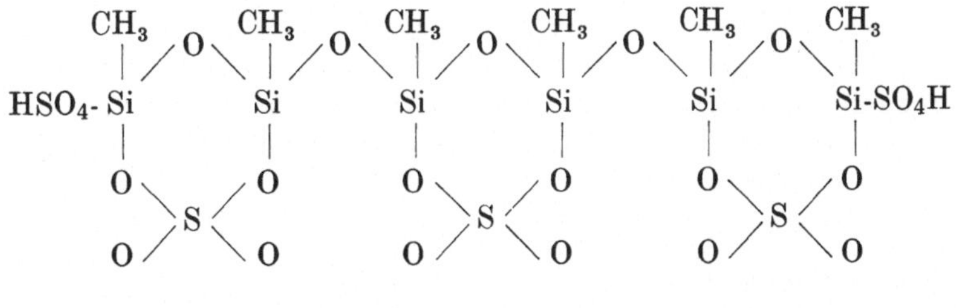

Salts of sulphostannic acid have been prepared[75,76], e.g. $M_2Sn(SO_4)_3$, by evaporation of mixtures of stannic oxide, metal sulphate and sulphuric acid. In sulphuric acid solution $H_2Sn(HSO_4)_6$ is found to exist:

[71] GILLESPIE, R. J., and E. A. ROBINSON: Canad. J. Chem. **40**, 784 (1962).
[72] SCHOTT, G., and H. U. KIBBEL: Z. anorg. allg. Chem. **314**, 104 (1962).
[73] GILLESPIE, R. J., and E. A. ROBINSON: Canad. J. Chem. **40**, 1009 (1962).
[74] FLOWERS, R. H., R. J. GILLESPIE, and E. A. ROBINSON: Canad. J. Chem. **41**, 2464 (1963).
[75] WEINLAND, R. F., and H. KUHL: Z. anorg. allg. Chem. **54**, 244 (1907).
[76] DRUCE, J. G. F.: Chem. News **128**, 33 (1924); C. A. **18**, 976.

$$(C_6H_5)_4Sn + 14\,H_2SO_4 = H_2Sn(HSO_4)_6 + 4\,C_6H_5SO_3H + 4\,H_3O^+ + 4\,HSO_4^-$$
$$Sn(CH_3COO)_4 + 10\,H_2SO_4 = H_2Sn(HSO_4)_6 + 4\,CH_3COOH_2^+ + 4\,HSO_4^-$$

Lead acetate reacts in an analogous manner,

$$Pb(OOCCH_3)_4 + 10\,H_2SO_4 = H_2Pb(HSO_4)_6 + 4\,CH_3COOH_2^+ + 4\,HSO_4^-$$

whilst vanadium(V) oxide is solvolysed by sulphuric acid to give $VO(HSO_4)_3$,

$$V_2O_5 + 9\,H_2SO_4 = 2\,VO(HSO_4)_3 + 3\,H_3O^+ + 3\,HSO_4^-$$

and polymeric species are simultaneously formed.

When K_2CrO_4 is reacted in oleum H_2CrSO_7 is formed:

$$K_2CrO_4 + 4\,H_2SO_4 \rightarrow HO-\underset{\underset{O}{|}}{\overset{\overset{O}{|}}{Cr}}-O-\underset{\underset{O}{|}}{\overset{\overset{O}{|}}{S}}-OH + 2\,K^+ + H_3O^+ + 3\,HSO_4^-$$

Potassium permanganate gives green solutions in sulphuric acid, but more concentrated solutions quickly become turbid, and this was interpreted as arising from the formation of permanganyl-hydrogen sulphate[77]:

$$KMnO_4 + 3\,H_2SO_4 = K^+ + H_3O^+ + MnO_3SO_4H + 2\,HSO_4^-$$

Unfortunately definite conclusions cannot be drawn from the available experimental evidence.

Sulphur dioxide gives slightly conducting solutions probably by accepting a proton,

$$SO_2 + H_2SO_4 \rightleftharpoons HSO_2^+ + HSO_4^-$$

and selenium dioxide gives partly polymeric protonated species.

Tellurium dioxide is, however, insoluble in sulphuric acid.

When iodine is reacted with iodic acid in sulphuric acid $(IO)_2SO_4$ is produced and is probably ionic containing the polymeric $[IO]_n^+$-ion[78]. When iodine is dissolved in a solution of iodosyl sulphate[79] in sulphuric acid, the I_3^+-ion appears to be present in the blue solution:

$$HIO_3 + 7\,I_2 + 8\,H_2SO_4 = 5\,I_3^+ + 3\,H_3O^+ + 8\,HSO_4^-$$

With further addition of iodine, I_5^+ seems to be formed:

$$I_3^+ + I_2 \rightleftharpoons I_5^+$$

Triphenylphosphine and triphenylarsine carbonyls such as $P(C_6H_5)_3Fe(CO)_4$ give stable yellow solutions in sulphuric acid[80]:

$$Fe(CO)_4P(C_6H_5)_3 + H_2SO_4 = [HFe(CO)_4P(C_6H_5)_3]^+ + HSO_4^-$$

[77] ROYER, J. L.: J. Inorg. Nucl. Chem. 17, 159 (1961).
[78] DASENT, W. E., and T. C. WADDINGTON: J. Chem. Soc. 1960, 3350.
[79] MASSON, I.: J. Chem. Soc. 1938, 1708.
[80] DAVISON, A., W. McFARLANE, L. PRATT, and G. WILKINSON: J. Chem. Soc. 1962, 3653.

With carbonyl-olefin metal complexes a proton is added to the olefinic ligand to give "carbonium"-ions[81]:

$$Fe(CO)_3 \cdots \text{⬡} + H^+ \rightarrow Fe(CO)_3 \cdots \text{⬡}(+)$$

Cyanocomplexes are protonated at the cyano-groups and not at the metal atom[82]

$$[Fe(phen)(CN)_4]^- + 4H^+ \rightarrow [Fe(phen)(CNH)_4]^{3+}$$

Another feature of coordination chemistry in sulphuric acid is due to the chelating properties of the sulphate group. This appears to occur in trimethylene sulphate[83]

$$H_2C \underset{CH_2-O}{\overset{CH_2-O}{<}} S \overset{O}{\underset{O}{<}}$$

and in certain sulphates, such as $(CH_3)_2Pb(SO_4)_3$, $K_3[Ce(SO_4)_3]$, and $Sb_2(SO_4)_3$, but direct evidence is lacking. For ethylgermanium sulphate a chelate structure has also been proposed[84]:

$$\underset{C_2H_5}{\overset{C_2H_5}{>}}Ge\underset{SO_4}{\overset{SO_4}{<}}Ge\underset{C_2H_5}{\overset{C_2H_5}{<}}$$

Several metathetical reactions have also been described, such as:

$$NaCl + AgNO_3 \rightleftharpoons AgCl + NaNO_3$$
$$HgSO_4 + 2NaCl + H_2SO_4 \rightleftharpoons HgCl_2 + 2NaHSO_4$$
$$CuSO_4 + 2NaCl + H_2SO_4 \rightleftharpoons CuCl_2 + 2NaHSO_4$$

6. Nitric Acid and Phosphoric Acid

Table 33. *Some Physical Properties of* HNO_3 *and* H_3PO_4

	HNO_3	H_3PO_4
Melting Point (°C)	−41	+42.35
Boiling Point (°C)	+86	
Density (g · cm⁻³)	1.547 (18°)	1.852 (45°)
Specific Conductivity (Ohm⁻¹ · cm⁻¹)	9.10⁻³ (20°)	8.84 · 10⁻² (45°)
Dielectric Constant at 20°	80	
Viscosity (Centipoise)		76.45 (45°)
Enthalpy of Vaporization (kcal · mole⁻¹)	7.25	

Nitric acid reacts with various donor compounds and shows no tendency to solvate metal cations. It is therefore reasonable to consider nitric acid as an acceptor solvent. Its strong proton donor properties are responsible for levelling the strengths of basic substances.

[81] DAVISON, A., W. MCFARLANE, L. PRATT, and G. WILKINSON: J. Chem. Soc. **1962**, 4821.
[82] SCHILT, A. A.: J. Amer. Chem. Soc. **85**, 904 (1963).
[83] BAKER, W., and F. B. FIELD: J. Chem. Soc. **1932**, 86.
[84] ANDERSON, H. H.: J. Amer. Chem. Soc. **72**, 194 (1950).

Nitric acid shows an appreciable electric conductivity in the pure liquid state[85]. This is partly due to autoprotolysis and partly due to a self-dehydration equilibrium:

$$2\,HNO_3 \rightleftharpoons H_2NO_3{}^+ + NO_3{}^- \qquad\qquad \text{autoprotolysis}$$
$$H_2NO_3{}^+ \rightleftharpoons H_2O + NO_2{}^+ \qquad\qquad \text{self-dehydration}$$
$$\underline{H_2O + HNO_3 \rightleftharpoons H_3O^+ + NO_3{}^-}$$
$$3\,HNO_3 \rightleftharpoons NO_2{}^+ + H_3O^+ + 2\,NO_3{}^-$$

Perchloric and sulphuric acids are extremely weak acids in this medium, and acetic acid acts as a base:

$$HNO_3 + CH_3COOH \rightleftharpoons NO_3{}^- + CH_3COOH_2{}^+$$

Solvates of alkali metal nitrates such as $KNO_3 \cdot HNO_3$ and $KNO_3 \cdot 2\,HNO_3$ have been described and it seems likely that the anions are solvated in these compounds[85-87].

Examples of reactions in nitric acid are[86, 87, 88]:

$$POCl_3 + HNO_3 \rightleftharpoons HPO_3 + Cl_2 + NOCl$$
$$[R_4N][NO_3] + UO_2(NO_3)_2 \rightleftharpoons [R_4N][UO_2(NO_3)_3]$$
$$2\,HClO_4 + UO_2(NO_3)_2 \rightleftharpoons [UO_2][ClO_4]_2 + 2\,HNO_3$$
$$[Cr(NH_3)_6](NO_3)_3 + 3\,HClO_4 \rightleftharpoons [Cr(NH_3)_6](ClO_4)_3 + 3\,HNO_3$$
$$[Co(NH_3)_6](NO_3)_3 + 3\,HClO_4 \rightleftharpoons [Co(NH_3)_6](ClO_4)_3 + 3\,HNO_3$$

The high electrical conductivity of fused phosphoric acid is regarded as due to extensive autoprotolysis

$$2\,H_3PO_4 \rightleftharpoons H_4PO_4{}^+ + H_2PO_4{}^-$$

followed by a slow reaction of self-dehydration[89, 90]:

$$2\,H_3PO_4 \rightleftharpoons H_3O^+ + H_3P_2O_7{}^-$$

The $H_2PO_4{}^-$ ion has so high a mobility that a proton-switch conduction mechanism is implied. The effect is even more striking than in H_2SO_4, since the very high viscosity of this liquid renders any contribution from normal ionic migration neglegible. The abnormal conductivity is removed by adding BF_3 when the complex $BF_3 \cdot H_3PO_4$ is formed[90, 91]. In this complex the BF_3 molecules occupy positions formerly available for hydrogen bonding:

$$\left[\begin{array}{c} HO \\ \\ HO \end{array} \!\! \diagup\!\!\!\!\!\diagdown \! P \! \diagup\!\!\!\!\!\diagdown\!\! \begin{array}{c} OH \\ \\ OH \end{array}\right]^+ \qquad \left[\begin{array}{c} HO \\ \\ HO \end{array} \!\! \diagup\!\!\!\!\!\diagdown \! P \! \diagup\!\!\!\!\!\diagdown\!\! \begin{array}{c} OBF_3 \\ \\ OBF_3 \end{array}\right]^-$$

[85] JANDER, G., and H. WENDT: Z. anorg. allg. Chem. **257**, 26 (1948).

[86] JANDER, G., and H. WENDT: Z. anorg. allg. Chem. **258**, 1 (1949).

[87] WENDT, H., and G. JANDER: Z. anorg. allg. Chem. **259**, 309 (1949).

[88] JANDER, G.: „Die Chemie in wasserähnlichen Lösungsmitteln", Springer-Verlag, Berlin-Göttingen-Heidelberg 1959.

[89] GREENWOOD, N. N., and A. THOMPSON: J. Chem. Soc. **1959**, 3485.

[90] GREENWOOD, N. N., and A. THOMPSON: Proc. Chem. Soc. **1958**, 352.

[91] GREENWOOD, N. N., and A. THOMPSON: J. Chem. Soc. **1959**, 3493.

7. Fluorosulphuric Acid, Chlorosulphuric Acid, Difluorophosphoric Acid and Disulphuric Acid

After the ionizing properties of fluorosulphuric acid were discovered[92] and more extensively studied[93-95], similar phenomena were observed in chlorosulphuric acid[96], difluorophosphoric acid[97] and disulphuric acid[98]. Their solvent properties resemble those of sulphuric acid.

Table 34. *Physical Properties of Fluorosulphuric Acid, Chlorosulphuric Acid, Difluorophosphoric Acid and Disulphuric Acid*

	HSO_3F	HSO_3Cl	HPO_2F_2	$H_2S_2O_7$
Melting Point (°C)	−89.0	−80.0	−96.5	+35.1
Boiling Point (°C)	+162.7	+152.0	116.0 (decomp.)	
Viscosity (Centipoise) at 25°	1.56	2.43	6.1	74.8
Density (g · cm⁻³) at 25°	1.7264	1.741	1.58	1.97
Spec. Conductivity at 25° (Ohm⁻¹cm⁻¹)	5.10^{-5}	4.10^{-4}	2.10^{-4}	4.10^{-3}
Dielectric Constant		60 ± 10		∼50

The following autoprotolysis equilibrium has been assumed for fluorosulphuric acid[92,93]

$$2\,HSO_3F \rightleftharpoons H_2SO_3F^+ + SO_3F^-$$

as well as the equilibria:

$$HSO_3F \rightleftharpoons SO_3 + HF$$
$$HF + HSO_3F \rightleftharpoons H_2F^+ + SO_3F^-$$
$$SO_3 + 2\,HSO_3F \rightleftharpoons H_2SO_3F^+ + S_2O_6F^-$$

The self-ionization is considerably smaller than that of sulphuric acid, as is indicated by a fairly sharp freezing point maximum in the system HF-SO₃ at a 1 : 1 molar ratio.

Alkali metal fluorosulphates act as bases in fluorosulphuric acid; their conductivities are all very similar and at any given concentration; they decrease slightly in the order $NH_4^+ > Rb^+ > K^+ > Na^+ > Li^+$. The abnormal mobility of both of the fluorosulphate ion and of the fluorosulphuric acidium ion $H_2SO_3F^+$ have been attributed to a proton transfer mechanism of conduction.

[92] WOOLF, A. A.: J. Chem. Soc. **1955**, 433.

[93] BARR, J., R. J. GILLESPIE, and R. C. THOMPSON: Inorg. Chem. **3**, 1149 (1964).

[94] THOMPSON, R. C., J. BARR, R. J. GILLESPIE, J. B. MILNE, and R. A. ROTHENBURY: Inorg. Chem. **4**, 1641 (1965).

[95] GILLESPIE, R. J., J. B. MILNE, and R. C. THOMPSON: Inorg. Chem. **5**, 468 (1965).

[96] ROBINSON, E. A.: presented at the Int. Conf. on Non-Aqueous Solvent Chemistry, McMaster University, Hamilton/Ontario (Canada) 1967.

[97] THOMPSON, R. C.: presented at the Int. Conf. on Non-Aqueous Solvent Chemistry, McMaster University, Hamilton/Ontario (Canada) 1967.

[98] MALHOTRA, C. H.: presented at the Int. Conf. on Non-Aqueous Solvent Chemistry, McMaster University, Hamilton/Ontario (Canada) 1967.

Basic behaviour is also exhibited by acetic acid, benzoic acid, nitromethane, nitrobenzene and other organic bases; the conductivities of their solutions are similar to those obtained for the metal fluorosulphates[93].

Basic behaviour of hydrogen fluoride in fluorosulphuric acid has been concluded from the results of electrolysis experiments[92] and of conductometric studies[93]. Arsenic(III) fluoride and antimony(III) fluoride are weak bases:

$$H^+$$

$$AsF_3 + HFSO_3 \rightleftharpoons HAsF_3^+ + FSO_3^-$$

while $S_2O_5F_2$ and $S_2O_6F_2$ are non-electrolytes.

WOOLF[92] found that potassium fluorosulphate could be titrated with antimony(V) fluoride and deduced that antimony(V) fluoride behaves as an acid in this solvent medium:

$$SbF_5 + SO_3F^- \rightleftharpoons F_5Sb(SO_3F)^-$$

The $F_5Sb(SO_3F)^-$ ion was shown to contain hexacoordinated antimony

$$\left[\begin{array}{c} F \quad\quad F \quad\; O \\ F \rightarrow Sb \leftarrow O - S - F \\ F \quad\quad F \quad\; O \end{array}\right]^-$$

and according to n.m.r. evidence appears also to exist as a dimer. Sulphur trioxide reacts in fluorosulphuric acid with antimony(V) fluoride to give compounds of the general formula[94]

$$[SbF_{5-n}(SO_3F)_{n+1}]^-$$

$HSbF_2(FSO_3)_4$ is a strong acid with a structure

$$\left[\begin{array}{c} FO_2SO \quad F \quad OSFO_2 \\ \quad\quad Sb \\ FO_2SO \quad F \quad OSFO_2 \end{array}\right]^-$$

which is in equilibrium with a dimeric species, while $HSbF_3(FSO_3)_3$ and $HSbF_4(FSO_3)_2$ are incompletely ionized.

The neutral fluorosulphates $SbF_{5-n}(SO_3F)_n$ are also known; n.m.r. measurements show that $SbF_2(SO_3F)_3$ is appreciably dissociated according to the equation[94]:

$$SbF_2(SO_3F)_3 \rightleftharpoons SbF_3(SO_3F)_2 + SO_3$$

The monofluorosulphate $SbF_4(SO_3F)$ has been prepared from antimony(V) fluoride and sulphur trioxide in the absence of any solvent[95].

Antimony(V) fluoride in fluorosulphuric acid becomes relatively more ionized as the concentration is increased; this has been explained by the formation of dimeric anions:

$$2 H[SbF_5(SO_3F)] \rightleftharpoons (SbF_5)_2SO_3F^- + H_2SO_3F^+$$

Boron(III) fluoride, gold(III) fluoride, tantalum(V) fluoride and platinum(IV) fluoride are further examples of weak acids in fluorosulphuric acid:

$$BF_3 + 3\,SO_3 + 2\,HFSO_3 \rightleftharpoons H_2SO_3F^+ + B(SO_3F)_4^-$$

but perchloric acid is nearly a non-electrolyte and potassium perchlorate is solvolysed to undissociated perchloric acid:

$$KClO_4 + HSO_3F \rightleftharpoons K^+ + SO_3F^- + HClO_4$$

Analogous behaviour is found in difluorophosphoric acid[97], in which boron(III) fluoride and antimony(V) fluoride behave as typical acids[97]. Fluorosulphuric acid protonates difluorophosphoric acid:

$$\overset{\displaystyle H^+}{\underset{\displaystyle }{\overbrace{HPO_2F_2}}} + HSO_3F \rightleftharpoons H_2PO_2F_2^+ + SO_3F^-$$

Ionization of carbonium ions has been found in chlorosulphuric acid[98]:

$$RCCl_3 + HSO_3Cl \rightleftharpoons RCCl_2^+ + SO_3Cl^- + HCl$$

In general chlorosulphuric acid is a good solvent for chlorides. Boron(III) chloride as well as antimony(V) chloride show acidic properties according to the reaction scheme:

$$BCl_3 + 5\,HSO_3Cl \rightleftharpoons H_2SO_3Cl^+ + B(SO_3Cl)_4^- + 3\,HCl$$

On the other hand selenium(IV) chloride, tellurium(IV) chloride and phosphorus(V) chloride act as bases

$$PCl_5 + HSO_3Cl \rightleftharpoons PCl_4^+ + SO_3Cl^- + HCl$$

and arsenic(III)chloride is a non-electrolyte in this solvent system.

Selenium oxychloride, selenium dioxide and tellurium dioxide are extensively ionized owing to protonation. From titanium(IV) chloride $TiCl_3(SO_3Cl)$ is obtained:

$$TiCl_4 + HSO_3Cl \rightarrow TiCl_3(SO_3Cl) + HCl$$

Fluorosulphuric acid has been used as a reaction medium and fluorinating agent to obtain various fluorides, such as permanganyl fluoride and perrhenyl fluoride[99]:

$$KMnO_4 + 2\,HSO_3F \rightarrow MnO_3F + KSO_3F + H_2SO_4$$

Other similar applications of this solvent medium include the preparation of selenyl fluoride[100] SeO_2F_2 and of pentafluorotelluric acid[101] F_5TeOH; the latter may be obtained from BaH_4TeO_4 and fluorosulphuric acid.

Recently anhydrous fluorosulphates of bivalent Mn, Fe, Co, Ni, Cu, Zn and Cd were prepared by displacement reactions in fluorosulphuric acid[102].

[99] ENGELBRECHT, A., and A. V. GROSSE: J. Amer. Chem. Soc. **76**, 2042 (1954).
[100] ENGELBRECHT, A., and B. STOLL: Z. anorg. allg. Chem. **292**, 20 (1957).
[101] ENGELBRECHT, A., and F. SLADKY: Mh. Chem. **96**, 159 (1965).
[102] WOOLF, A. A.: J. Chem. Soc. **1967**, (A), 355.

Chapter V

Proton-free Acceptor Solvents

1. Covalent Oxides

A. Liquid Sulphur Dioxide

The solvent properties of sulphur dioxide were discovered some time ago[1-6]; several reviews are available[7-11].

Table 35. *Some Physical Properties of* SO_2

Melting Point (°C)	-95.5
Boiling Point (°C)	-10.0
Density at $-12°$ (g · cm^{-3})	1.46
Dielectric Constant at 0°	15.4
Dipole Moment (Debye)	1.62
Specific Conductivity at $-10°$ (Ohm^{-1} · cm^{-1}) .	3.10^{-8}
Enthalpy of Vaporization (kcal · mole^{-1})	5.96

Apart from many covalent substances sulphur dioxide dissolves metal sulphites[7] (with formation of disulphites) as well as tetraalkylammonium halides, alkali halides and pseudohalides to some extent, and numerous covalent compounds.

[1] WALDEN, P.: Ber. dtsch. chem. Ges. **35**, 2018 (1902).

[2] WALDEN, P., and M. CENTNERSZWER: Ber. dtsch. chem. Ges. **32**, 2862 (1899).

[3] WALDEN, P., and M. CENTNERSZWER: Z. physik. Chem. **39**, 513 (1902).

[4] WALDEN, P., and M. CENTNERSZWER: Z. anorg. Chem. **30**, 145 (1902).

[5] WALDEN, P., and M. CENTNERSZWER: Z. anorg. Chem. **30**, 179 (1902).

[6] WALDEN, P., and M. CENTNERSZWER: Z. physik. Chem. **42**, 432 (1903).

[7] JANDER, G.: „Die Chemie in wasserähnlichen Lösungsmitteln", Springer-Verlag, Berlin-Göttingen-Heidelberg 1949.

[8] AUDRIETH, L. F., and J. KLEINBERG: "Non-Aqueous Solvents," Wiley, New York 1953.

[9] ELVING, P. J., and J. M. MARKOWITZ: J. Chem. Educ. **37**, 75 (1960).

[10] LICHTIN, N. N.: "Progress in Physical Organic Chemistry," Vol. I, Wiley, New York 1963.

[10a] KARCHER, W., and H. HECHT: Vol. III, Part 2 in "Chemistry in Nonaqueous Ionizing Solvents," Ed. G. JANDER, H. SPANDAU, and C. C. ADDISON, Vieweg, Braunschweig and Pergamon Press, London and New York 1967.

[11] WADDINGTON, T. C.: "Non-Aqueous Solvent Systems," Chapter 6, Academic Press, London-New York 1965.

A few solvates are known to be formed with covalent acceptor compounds apparently due to a certain amount of donor properties. Examples are: $F_3B \cdot SO_2$[12], $Cl_5Sb \cdot SO_2$[13], $Br_4Sn \cdot SO_2$, $(TiCl_4)_2SO_2$[14], $Cl_4Zr \cdot SO_2$[15], $Cl_3Al \cdot SO_2$[16,17]. SO_2 is known to be coordinated through the sulphur atom in complex compounds of class (b) metals[18-20].

The acceptor functions are shown by the reaction with OH^--ions,

$$SO_2 + OH^- \rightleftharpoons HSO_3^-,$$

by the solvation of the sulphite ion,

$$\underset{\text{donor}}{SO_3^{--}} + \underset{\text{acceptor}}{SO_2} \rightleftharpoons S_2O_5^{--}$$

and by its reactions with amines[21, 22] yielding adducts,

$$\underset{\text{donor}}{(C_2H_5)_3N} + \underset{\text{acceptor}}{SO_2} \rightleftharpoons (C_2H_5)_3N \cdot SO_2$$

which was incorrectly reported as $\{[(C_2H_5)_3N\}_2SO][SO_3]$ by JANDER and WICKERT[23]. The structure of the adduct with triethylamine oxide is[24]

$$(C_2H_5)_3N\!-\!S\!\!\begin{array}{c}O\\\diagdown\\O\end{array}$$

donor   acceptor

The acceptor properties of sulphur dioxide are also responsible for the interaction with numerous weak donors such as benzene and derivatives[25], olefines[26, 27], alcohols and for the ionization of triphenylcarbonium salts[10], which remain un-ionized in donor solvents and in inert solvents:

$$\underset{\text{donor}}{(C_6H_5)_3CCl} + \underset{\text{acceptor}}{SO_2} \rightleftharpoons (C_6H_5)_3C^+ + SO_2Cl^-$$

The existence of a self ionization according to

$$2SO_2 \rightleftharpoons SO^{++} + SO_3^{--}$$

was postulated by JANDER[7] and served its purpose as many new reactions were discovered. It is, however, apparent that such self ionization does not take place to any measurable extent. Thionyl halides, which were considered as solvoacids, give practically non-conducting solutions, which do not show any exchange with

[12] BROTH, H. S., and D. R. MARTIN: J. Amer. Chem. Soc. 64, 2198 (1942).
[13] AYNSLEY, E. E., R. D. PEACOCK, and P. L. ROBINSON: Chem. Ind. 1951, 1117.
[14] BOND, P. A., and W. E. BELTON: J. Amer. Chem. Soc. 67, 1691 (1945).
[15] BOND, P. A., and W. R. STEPHENS: J. Amer. Chem. Soc. 51, 2910 (1929).
[16] BAUDE, E.: Ann. Chem. Phys. 1, 8 (1904).
[17] BURG, A. B., and J. H. BICKERTON: J. Amer. Chem. Soc. 67, 2261 (1945).
[18] VOGT, L. H., J. L. KATZ, and S. E. WIBERLEY: Inorg. Chem. 4, 1157 (1965).
[19] LaPLACA, S. J., and J. A. IBERS: Inorg. Chem. 5, 405 (1966).
[20] LEVISON, J. J., and S. D. ROBINSON: Chem. Comm. 1967, 198.
[21] BATEMAN, L. C., E. D. HUGHES, and C. K. INGOLD: J. Chem. Soc. 1944, 243.
[22] BALEJ, J., and A. REGNER: Chem. Listý 50, 374, 381 (1956).
[23] JANDER, G., and K. WICKERT: Z. physik. Chem. A 178, 57 (1936).
[24] LECHER, H. Z., and W. B. HARDY: J. Amer. Chem. Soc. 70, 3789 (1948).
[25] ANDREWS, L. J., and R. M. KEEFER: J. Amer. Chem. Soc. 73, 4169 (1951).
[26] BOOTH, D., F. S. DAINTON, and K. J. IVIN: Trans. Farad. Soc. 55, 1293 (1959).
[27] DE MAINE, P. A. D.: J. Chem. Phys. 26, 1036 (1957).

the solvent molecules[28-36]. On the other hand rapid exchange is observed with sulphites because of the acceptor properties of the solvent molecules towards these ions:

$$SO_2 + SO_3^{--} \rightleftharpoons S_2O_5^{--}$$

Exchange is further found in the system triethylamine-sulphur dioxide in the presence of thionyl chloride, where the formation of a transition species was postulated such as

The reaction

$$[(CH_3)_4N]_2SO_3 + SOCl_2 = 2(CH_3)_4NCl + 2SO_2$$

may be interpreted without the assumption of the self-ionization of liquid sulphur dioxide[11].

The formation of compounds containing complex anions is expected to occur easily in an acceptor solvent such as sulphur dioxide. The formation of K_3SbCl_6 and $KSbCl_6$ has been reported[37]. Reactions between nitrosyl chloride, acetyl chloride or benzoyl chloride and antimony(V) chloride lead in liquid sulphur dioxide to the respective chloro-complexes, namely $NO[SbCl_6]$, $[CH_3CO]SbCl_6$ and $[C_6H_5CO][SbCl_6]$[38-41], but the latter has recently been shown to be

C_6H_5C—$SbCl_5$ in the solid state[42]. The solution of nitrosyl hexachloro-antimonate can be used to prepare a number of other nitrosonium compounds by metathetical reactions. For example $NOPF_6$ is precipitated by addition of tetramethylammoniumhexafluorophosphate[43] and in an analogous manner

[28] NORRIS, T. H.: J. Phys. Chem. **63**, 383 (1959).

[29] HUSTON, J. L.: J. Amer. Chem. Soc. **73**, 3049 (1951).

[30] HUSTON, J. L.: J. Phys. Chem. **63**, 389 (1959).

[31] JOHNSON, R. E., T. H. NORRIS, and J. L. HUSTON: J. Amer. Chem. Soc. **73**, 3052 (1951).

[32] MASTERS, B. J., and T. H. NORRIS: J. Amer. Chem. Soc. **77**, 1346 (1955).

[33] JOHNSON, R. E., and T. H. NORRIS: J. Amer. Chem. Soc. **79**, 1584 (1957).

[34] BURGE, D. E., H. FREUND, and T. H. NORRIS: J. Phys. Chem. **63**, 1969 (1959).

[35] HERBER, R. H., T. H. NORRIS, and J. L. HUSTON: J. Amer. Chem. Soc. **76**, 2015 (1954).

[36] MASTERS, B. J., N. D. POTTER, D. R. ASHER, and T. H. NORRIS: J. Amer. Chem. Soc. **78**, 4252 (1956).

[37] JANDER, G., and H. IMMIG: Z. anorg. allg. Chem. **233**, 295 (1937).

[38] SEEL, F.: Z. anorg. Chem. **250**, 331 (1943).

[39] SEEL, F.: Z. anorg. Chem. **252**, 24 (1943).

[40] SEEL, F., and H. BAUER: Z. Naturf. **2b**, 397 (1947).

[41] SEEL, F., J. NOGRADI, and R. ROSSE: Z. anorg. allg. Chem. **269**, 197 (1952).

[42] WEISS, R., and B. CHEVRIER: Chem. Comm. **1967**, 145.

[43] SEEL, F., and TH. GÖSSL: Z. anorg. allg. Chem. **263**, 253 (1950).

nitrosonium nitroprusside $[NO]_2[Fe(NO)(CN)_5]$ can be obtained[44, 45]. Indications are also available for the formation of potassium tetrachloroborate from potassium acetate and boron(III) chloride in liquid sulphur dioxide[34].

Nitrosyl fluoride reacts with sulphur dioxide and the compound $SO_2 \cdot NOF$ can be isolated[46]. It reacts with fluorides to form the respective nitrosonium fluorocomplexes, such as the fluoroborate or the fluorosilicate[47].

Boron(III) chloride is converted into the fluoride[41]:

$$BCl_3 + 3SO_2 \cdot NOF = BF_3 + 3NOCl + 3SO_2$$

and the chlorides of phosphorus, arsenic and antimony yield nitrosonium salts of the hexafluoroacids, e.g.

$$AsCl_3 + 6SO_2 \cdot NOF = NOAsF_6 + 2NO + 3NOCl + 6SO_2$$

With sulphur trioxide nitrosonium fluorosulphate $NO(SO_3F)$ is formed[47].

Another example for the acceptor properties of sulphur dioxide is the formation of alkalifluorosulphinates MSO_2F which are obtained from solutions of alkali fluorides in liquid sulphur dioxide[48, 49]

$$\underset{\text{donor}}{KF} \ + \ \underset{\text{acceptor}}{SO_2} \ \rightleftharpoons \ K^+SO_2F^-$$

These are isomorphous with the corresponding isosteric chlorates[50] and have been shown to fluorinate chlorides or oxychlorides, such as thionyl chloride or arsenic(III) chloride, to the respective fluorides[50].

Furthermore complex iodides of cadmium and mercury seem to be formed[5]. It is very likely that sulphur dioxide will be a useful medium for the preparation of a number of bromo- and iodo-complexes which are not obtained from aqueous solutions. It is, however, neccessary to exclude traces of moisture, which in the presence of iodide ions might promote the occurrance of undesired redox reactions[10]. It appears that sulphur dioxide offers many more possibilities as a medium for the preparation of various inorganic complex compounds.

Various solvolysis reactions are useful for the preparation of oxyhalides[7,8,51,52]:

$$PCl_5 + SO_2 \rightarrow POCl_3 + SOCl_2$$
$$PBr_5 + SO_2 \rightarrow POBr_3 + SOCl_2$$
$$NbCl_5 + SO_2 \rightarrow NbOCl_3 + SOCl_2$$
$$WCl_6 + SO_2 \rightarrow WOCl_4 + SOCl_2$$
$$2UCl_5 + 2SO_2 \rightarrow UO_2Cl_2 + UCl_4 + 2SOCl_2$$
$$MoCl_5 + SO_2 \rightarrow MoOCl_3 + SOCl_2$$
$$WBr_5 + SO_2 \rightarrow WOBr_4 + \text{byproducts.}$$

The last reaction in this group is unique since an increase in the metal oxidation state occurs.

[44] SEEL, F.: Z. anorg. allg. Chem. **261**, 81 (1950).
[45] SEEL, F., and N. H. WALASSIS: Z. anorg. allg. Chem. **261**, 85 (1950).
[46] SEEL, F., and H. MEIER: Z. anorg. allg. Chem. **274**, 196 (1953).
[47] SEEL, F., and H. MASSAT: Z. anorg. allg. Chem. **280**, 185 (1955).
[48] SEEL, F., and H. MEIER: Z. anorg. allg. Chem. **274**, 202 (1953).
[49] SEEL, F., and L. RIEHL: Z. anorg. allg. Chem. **282**, 293 (1955).
[50] SEEL, F., H. JONAS, L. RIEHL, and J. LANGER: Angew. Chem. **67**, 32 (1955).
[51] FOWLES, G. W. A., and J. L. FROST: J. Chem. Soc. (A) **1967**, 671.
[52] EDWARDS, D. A.: J. Inorg. Nucl. Chem. **25**, 1198 (1963).

Metathetical reactions may be carried out readily:

$$Cs_2SO_3 + SOCl_2 \rightarrow 2\,CsCl + 2\,SO_2$$
$$2\,Ag(CH_3COO) + SOCl_2 \rightarrow 2\,AgCl + SO(CH_3COO)_2$$
$$2\,NH_4(CH_3COO) + SOCl_2 \rightarrow 2\,NH_4Cl + SO(CH_3COO)_2$$
$$2\,NH_4SCN + SOCl_2 \rightarrow 2\,NH_4Cl + SO(SCN)_2$$

Thionyl thiocyanate appears to be present in dilute solutions in liquid sulphur dioxide[53] at $-15°$, but on concentrating the solution polymerization to amorphous polycyanogen takes place. The solution of thionyl thiocyanate[53] reacts with potassium sulphite to give potassium thiocyanate[53]:

$$K_2SO_3 + SO(SCN)_2 = 2\,KSCN + 2\,SO_2$$

Other displacement reactions[37, 54-56] have been considered as "amphoteric reactions":

$$2\,AlCl_3 + 3\,[(CH_3)_4N]_2SO_3 = Al_2(SO_3)_3 + 6\,(CH_3)_4NCl$$
$$Al_2(SO_3)_3 + 3\,[(CH_3)_4N]_2SO_3 = 2\,[(CH_3)_4N]_3Al(SO_3)_3$$

On addition of thionyl chloride the sulphito-complex is destroyed and insoluble aluminium sulphite is precipitated:

$$2\,[(CH_3)_4N]_3Al(SO_3)_3 + 3\,SOCl_2 = Al_2(SO_3)_3 + 6\,(CH_3)_4NCl + 6\,SO_2$$

The reaction

$$2\,KBr + SOCl_2 = 2\,KCl + SOBr_2$$

may be used for the preparation of thionyl bromide[7].

Many other reactions in liquid sulphur dioxide, such as the occurrence of redox-reactions, the sulphonation of aromatic compounds or the possibility of carrying out FRIEDEL-CRAFTS reactions have been reviewed elsewhere[57].

B. Liquid Dinitrogen Tetroxide*

Table 36. *Some Physical Properties of Dinitrogen Tetroxide*

Melting Point (°C)	-12.3
Boiling Point (°C)	$+21.3$
Density at 0° (g · cm^{-3})	1.49
Dielectric Constant at 0°	2.42
Specific Conductivity at 17° (Ohm^{-1}cm^{-1}) ...	2.10^{-13}

As an acceptor solvent dinitrogen tetroxide forms molecular addition compounds with various donor molecules. Amines such as triethylamine, pyridine and quinoline give crystalline compounds which appear to be ionic[58], but sterically hindered complexes are not formed. Thus 2,6-lutidine and 2-methylquinoline do not give adducts with dinitrogen tetroxide.

[53] JANDER, G., and D. ULLMANN: Z. anorg. allg. Chem. **230**, 405 (1937).
[54] JANDER, G.: Naturwiss. **26**, 779 (1938).
[55] JANDER, G., and H. HECHT: Z. anorg. Chem. **250**, 287 (1943).
[56] JANDER, G., H. WENDT, and H. HECHT: Ber. dtsch. chem. Ges. **77**, 698 (1944).
[57] GUTMANN, V., and H. SPANDAU: Angew. Chem. **62**, 321 (1950).
[58] DAVENPORT, D., H. BURKHARDT, and H. H. SISLER: J. Amer. Chem. Soc. **75**, 4175 (1953).
* Recent review: C. C. ADDISON: Vol. III, Part 1 in "Chemistry in Nonaqueous Ionizing Solvents," Ed. G. JANDER, H. SPANDAU, and C. C. ADDISON, Vieweg Braunschweig, and Pergamon Press, London and New York, 1967.

Ethers (dioxane, tetrahydrofurane, diethylether), and other O-containing donor molecules, such as dimethyl sulphoxide, ketones and nitriles give compounds with molecular structures which have been considered as partially ionic in the solid state[59].

Weak molecular adducts are formed with benzene, mesitylene and nitrobenzene[59, 60]. Towards strong acceptors it may also act as weak donor, as is shown by its reaction with boron(III) fluoride to give

$$NO_2^+ \left[N \begin{matrix} \diagup OBF_3 \\ \diagdown OBF_3 \end{matrix} \right]^-$$

For the pure liquid a self-ionization equilibrium has been assumed to exist[61, 62] according to the equation:

$$N_2O_4 \rightleftharpoons NO^+ + NO_3^-$$

with an ionic product below 10^{-25}. Correspondingly nitrosonium compounds will act as solvo-acids and nitrates as solvo-bases. The reaction

$$\underset{\substack{\text{solvo-}\\\text{acid}}}{NOCl} \quad + \quad \underset{\substack{\text{solvo-}\\\text{base}}}{AgNO_3} \quad \rightleftharpoons \quad AgCl \quad + \quad N_2O_4$$

may be considered a neutralization reaction.

Although dinitrogen tetroxide is a symmetrical molecule

$$\begin{matrix} O \diagdown & & \diagup O \\ & N{-}N & \\ O \diagup & & \diagdown O \end{matrix}$$

it reacts as nitrosonium nitrate $[NO^+][NO_3^-]$.

Dinitrogen tetroxide is a poor solvent for ionic compounds. Covalent compounds are frequently soluble as are donor compounds which can react with the solvent.

Various compounds are solvolysed to give the corresponding nitrates[49, 54]:

$$KCl + N_2O_4 \rightleftharpoons NOCl + KNO_3$$
$$Mg(ClO_4)_2 + 2 N_2O_4 \rightleftharpoons Mg(NO_3)_2 + 2 NOClO_4$$
$$Li_2CO_3 + 2 N_2O_4 \rightleftharpoons 2 LiNO_3 + N_2O_3 + CO_2$$

Many metals are dissolved by liquid dinitrogen tetroxide with evolution of nitric oxide and formation of the respective metal nitrate[62-65].

$$Na + N_2O_4 \rightarrow NO\uparrow + NaNO_3$$

[59] ADDISON, C. C., and J. SHELDON: J. Chem. Soc. **1956**, 1941.

[60] SISLER, H. H.: "Chemistry in Non-Aqueous Solvents," Reinhold Publ., New York-London 1964.

[61] ADDISON, C. C., and R. THOMPSON: J. Chem. Soc. **1950**, S, 211.

[62] ADDISON, C. C., and R. THOMPSON: J. Chem. Soc. **1950**, S, 218.

[63] SISLER, H. H.: J. Chem. Educ. **34**, 555 (1957).

[64] GRAY, P.: Monograph Nr. 4, The Royal Institute of Chemistry, London 1958.

[65] ADDISON, C. C., and J. LEWIS: Quart. Revs. **9**, 115 (1955).

The reactivity of liquid dinitrogen tetroxide towards metals is increased by addition of nitrosyl chloride, which is considered a solvo-acid in this system. Thus zinc, iron and tin (M) are easily dissolved:

$$M + 2\,NOCl \xrightarrow{\text{liq. } N_2O_4} MCl_2 + 2\,NO\uparrow$$

In aqueous chemistry it has been observed that metals such as aluminium or zinc, which will not dissolve in water can be dissolved in basic solutions. Likewise the presence of diethylammonium nitrate in dinitrogen tetroxide will promote the solubility of zinc in this solvent:

$$Zn + n[(C_2H_5)_2NH_2]NO_3 + 2\,n\,N_2O_4 \rightarrow [(C_2H_5)_2NH_2]_n[Zn(NO_3)_{n+2}] + 2\,NO$$

Complex nitrates may also be obtained according to the reaction scheme:

$$R_4NCl + 4\,N_2O_4 + FeCl_3 \rightarrow R_4N[Fe(NO_3)_4] + 4\,NOCl$$

The rates of reaction of copper, zinc and uranium with dinitrogen tetroxide are greatly increased in the presence of acetonitrile[66, 67]. Reactions of this type lead to the formation of acetonitrile complexes of the corresponding nitrate, which may loose coordinated acetonitrile by pumping *in vacuo* or heating. Reactions in such mixed acceptor solvent-donor solvent systems may provide unusual solvent properties and it is desirable that such systems should be investigated in considerable detail.

2. Covalent Fluorides

Fluorocomplexes may be prepared in various covalent fluorides as solvents, such as liquid hydrogen fluoride, bromine(III) fluoride, chlorine(III) fluoride, iodine(V) fluoride, arsenic(III) fluoride or even the pentafluorides of vanadium, niobium and tantalum. All of them seem to be associated in the liquid states due to fluorine-bonding and splitting of such bonds may lead to fluoride ion transfer processes between solvent molecules (autofluoridolysis)[68, 69].

$$
\begin{aligned}
2\,HF + HF &\rightleftharpoons H_2F^+ + HF_2^- \\
BrF_3 + BrF_3 &\rightleftharpoons BrF_2^+ + BrF_4^- \\
ClF_3 + ClF_3 &\rightleftharpoons ClF_2^+ + ClF_4^- \\
IF_5 + IF_5 &\rightleftharpoons IF_4^+ + IF_6^- \\
AsF_3 + AsF_3 &\rightleftharpoons AsF_2^+ + AsF_4^- \\
VF_5 + VF_5 &\rightleftharpoons VF_4^+ + VF_6^- \\
NbF_5 + NbF_5 &\rightleftharpoons NbF_4^+ + NbF_6^- \\
TaF_5 + TaF_5 &\rightleftharpoons TaF_4^+ + TaF_6^-
\end{aligned}
$$

[66] ADDISON, C. C., J. C. SHELDON, and N. HODGE: J. Chem. Soc. **1956**, 3906.

[67] ADDISON, C. C., and N. LOGAN: "Preparative Inorganic Reactions," Ed. W. L. JOLLY, Vol. I, p. 141.

[68] GUTMANN, V., and I. LINDQVIST: Z. physik. Chem. **203**, 250 (1954).

[69] GUTMANN, V.: Svensk Kem. Tidskr. **68**, 1 (1955).

A. Bromine(III) fluoride and Chlorine(III) fluoride

Bromine(III) fluoride is a pale yellow liquid at room temperature, while chlorine(III) fluoride has to be cooled below 12° to be available in the liquid state at atmospheric pressure. Both are highly reactive and explode in the presence of water and most organic matter, react with asbestos with incandescence, but can be manipulated in quartz apparatus because the reactions with silica in this form are slow. Several reviews are available[69-74].

Table 37. *Some Physical Properties of* BrF_3 *and* ClF_3

Property	ClF_3	BrF_3
Melting Point (°C)	−83.0	+9.0
Boiling Point (°C)	+12.0	+126.0
Density (g · cm⁻³)	1.84 (12°)	2.80 (25°)
Dielectric Constant	4.6 (12°)	
Specific Conductivity (Ohm⁻¹ · cm⁻¹)	6.10^{-9} (0°)	8.10^{-3} (25°).
Heat of Vaporization (kcal · mole⁻¹)	6.6	10.2
Viscosity (Centipoise)	0.48 (12°)	2.22 (25°)

The molecules of ClF_3 and BrF_3 are T-shaped (as part of a trigonal bipyramid) and there is probably appreciable association in the pure liquids due to fluorine bridging

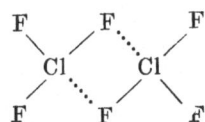

N.m.r. evidence suggests that chlorine trifluoride may be in the liquid state in equilibrium with a dimeric species[75]

This type of structure is known for iodine(III) chloride, which has been described to a first approximation in terms of an ionic structure and, due to the differences in bond lengths that were observed[76], with resonance between the two forms:

[70] GUTMANN, V.: Angew. Chem. **62**, 312 (1950).

[71] LEECH, H. R. in Mellor's "Comprehensive Treatise on Inorganic and Theoretical Chemistry", Suppl. II, Part I (1956).

[72] RYSS, I. G.: "The Chemistry of Fluorine and its Inorganic Compounds," US Atomic Energy Commission, 1960.

[73] GUTMANN, V.: Quart. Revs. **10**, 451 (1956).

[74] SHARPE, A. G.: Chapter 7 in "Non-Aqueous Solvent Systems," Ed. T. C. WADDINGTON, Academic Press, London-New York 1965.

[75] MUETTERTIES, E. L., and W. D. PHILLIPS: J. Amer. Chem. Soc. **79**, 322 (1957).

[76] BOSWIJK, K. H., and E. H. WIEBENGA: Acta Cryst. **7**, 417 (1954).

$$\left[\begin{array}{c}Cl \\ Cl\end{array}\!\!>\!\!I\right]^{+} \quad \left[\begin{array}{c}Cl \\ Cl\end{array}\!\!>\!\!I\!\!<\!\!\begin{array}{c}Cl \\ Cl\end{array}\right]^{-} \leftrightarrow \left[\begin{array}{c}Cl \\ Cl\end{array}\!\!>\!\!I\!\!<\!\!\begin{array}{c}Cl \\ Cl\end{array}\right]^{-} \quad \left[I\!\!<\!\!\begin{array}{c}Cl \\ Cl\end{array}\right]^{+}$$

The formation of $[BrF_2]^+$ and $[BrF_4]^-$-ions may also be explained by assuming a fluorine bridged structure in the liquid.

The existence of BrF_4^- and ClF_4^--ions has been established and their structures were found to be planar[77]. The compounds $NaBrF_4$, $KBrF_4$, $AgBrF_4$ and $Ba(BrF_4)_2$[78] are probably ionic just as is accepted for $KClF_4$, $RbClF_4$ and $Ca(ClF_4)_2$. The latter may be obtained from the respective fluorides and chlorine(III) fluoride[79] or by fluorination of the alkali chlorides[79]; they were assigned wrong formulas by Bode and Klesper[80].

A cationic species has been assumed to be present in a number of compounds formed from solvent molecules and various acceptor fluorides, such as antimony(V) fluoride. By analogy with the structure of ICl_3SbCl_5 which contains chlorine bridges[81] with still discrete ICl_2-groups

fluorine bridging may be assumed in the compounds under consideration with BrF_2-groups.

Compounds of this type are BrF_2SbF_6, $(BrF_2)SnF_6$[82], BrF_2BiF_6, BrF_2NbF_6, BrF_2TaF_6[83], $(BrF_2)_2OsF_6$[84], BrF_2AuF_4[85] and BrF_2SO_3F[86], which are good conductors in liquid bromine(III) fluoride. Likewise the compounds ClF_2SbF_6[87] and ClF_2BF_4[88] give conducting solutions in liquid chlorine(III) fluoride.

The solubilities of fluorides in bromine(III) fluoride are similar to those in liquid hydrogen fluoride, but few observations have been made in liquid chlorine(III) fluoride. Alkali fluorides show good solubilities with the formation of solvated anions, fluorides of the alkaline earth metals have limited solubilities with the exception of barium fluoride. Most of the other ionic fluorides are sparingly soluble or insoluble, since solvation of the metal ions would require a

[77] Sly, W. G., and R. E. Marsh: Acta Cryst. 10, 378 (1957).

[78] Sharpe, A. G., and H. J. Eméleus: J. Chem. Soc. 1948, 2135.

[79] Asprey, L. B., J. L. Margrave, and M. E. Silberthorn: J. Amer. Chem. Soc. 83, 2955 (1961).

[80] Bode, H., and E. Klesper: Z. anorg. allg. Chem. 267, 97 (1951).

[81] Vonk, C. G., and E. H. Wiebenga: Acta Cryst. 10, 378 (1957).

[82] Woolf, A. A., and H. J. Eméleus: J. Chem. Soc. 1949, 2865.

[83] Gutmann, V., and H. J. Eméleus: J. Chem. Soc. 1950, 1076.

[84] Sharpe, A. G.: J. Chem. Soc. 1950, 3444.

[85] Sharpe, A. G.: J. Chem. Soc. 1949, 2901.

[86] Woolf, A. A., and H. J. Eméleus: J. Chem. Soc. 1950, 1050.

[87] Seel, F., and O. Detmar: Angew. Chem. 70, 163, 470 (1958).

[88] Şelig, H., and J. Shamir: Inorg. Chem. 3, 294 (1964).

donor solvent. On the other hand covalent fluorides show reasonable or high solubilities and frequently give conducting solutions[78,89]. Several fluorides in which the metal has a low oxidation number, are converted into the respective highest fluorides by bromine- or chlorine trifluorides.

Fluorides of the alkali and alkaline earth metals, as well as silver, nitrosyl and nitryl fluorides act as bases since the solvent molecules accept the fluoride ions formed in their solutions:

$$F^-$$

$$\begin{array}{ccc} KF & + & BrF_3 & \rightleftharpoons & K^+BrF_4^- \\ KF & + & ClF_3 & \rightleftharpoons & K^+ClF_4^- \end{array}$$

donor   acceptor

Fluorides of other elements may accept fluoride ions and thus may be considered as acids in the liquid halogen fluorides.

$$BrF_3SbF_5 \rightleftharpoons BrF_2^+ + SbF_6^-$$

Other acids in bromine(III) fluoride are the fluorides of boron, gold(III), silicon, germanium, tin(IV), titanium(IV), phosphorus(V), arsenic(V), bismuth(V), vanadium(V), niobium(V), tantalum(V), ruthenium(V), platinum(V) as well as hydrogen fluoride and sulphur trioxide:

$$F^-$$

$$\begin{array}{ccc} BrF_3 & + & HF & \rightleftharpoons & BrF_2^+ & + & HF_2^- \\ BrF_3 & + & SO_3 & \rightleftharpoons & BrF_2^+ & + & SO_3F^- \end{array}$$

Complex fluorides are formed by neutralization reactions[82,90]:

$$BrF_2SbF_6 + AgBrF_4 \rightleftharpoons AgSbF_6 + 2\,BrF_3$$
$$VF_5 + KBrF_4 \rightleftharpoons KVF_6 + BrF_3$$

Bromine(III) fluoride fluorinates everything which dissolves in it. Numerous oxides, halides and salts of oxyacids are converted into fluorides. Carbonates, nitrates and iodates are usually completely converted into the fluorides, while some of the metal oxides, such as BeO, MgO, ZnO or Al_2O_3 are only partly fluorinated by liquid bromine(III) fluoride. Sodium vanadate yields a mixture of tetrafluorooxovanadate and hexafluorovanadate[91], potassium or silver dichromates give $KCrOF_4$ and $AgCrOF_4$ resp.[91]. Potassium permanganate is converted into $KMnF_5$, potassium metaphosphate into potassium hexafluorophosphate and sodium borate into the tetrafluoroborate[92].

Most of the metals react vigorously with either bromine(III) fluoride or chlorine(III) fluoride. Examples are the reactions with potassium, silver, barium, tin, vanadium, niobium, tantalum, molybdenum and tungsten.

$$3\,Ag + 4\,BrF_3 \rightarrow 3\,AgBrF_4 + \tfrac{1}{2}\,Br_2$$
$$3\,V + 5\,BrF_3 \rightarrow 3\,VF_5 + {}^5\!/_2\,Br$$

[89] SHEFT, I., H. H. HYMAN, and J. J. KATZ: J. Amer. Chem. Soc. **75**, 5221 (1953).
[90] EMELÉUS, H. J., and V. GUTMANN: J. Chem. Soc. **1949**, 2979.
[91] SHARPE, A. G., and A. A. WOOLF: J. Chem. Soc. **1951**, 798.
[92] EMELÉUS, H. J., and A. A. WOOLF: J. Chem. Soc. **1950**, 164.

Complex fluorides are easily prepared by making use of both the fluorinating and ionizing properties of bromine(III) fluoride. Compounds which will yield the acidic and basic fluorides by fluorination are mixed and allowed to react with excess bromine(III) fluoride. A mixture of equivalent amounts of potassium chloride and antimony(III) fluoride, for example, may be used to obtain potassium

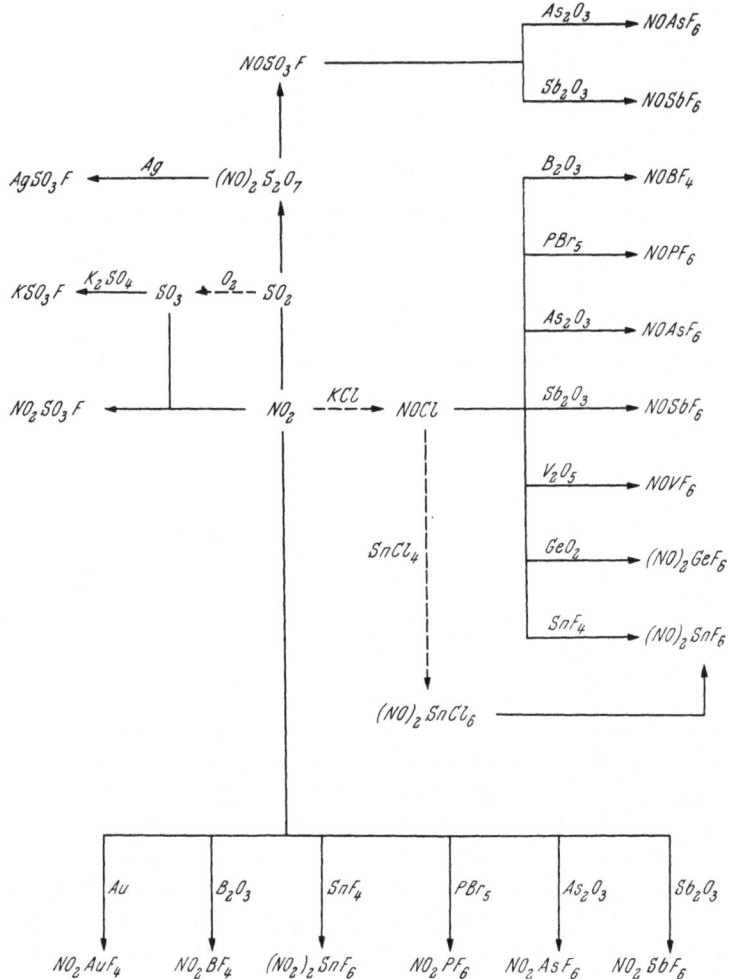

Fig. 9. Flow sheet for some reactions in liquid BrF_3

hexafluoroantimonate, while a mixture of silver and gold in equimolar properties will yield the insoluble $AgAuF_4$ by treatment with bromine(III)fluoride[85]. Hexafluorovanadates[90], hexafluororuthenates[93], pentafluoromanganates[91], complex oxyfluorides of rhenium[94] and a number of nitrosonium and nitronium compounds (see Fig. 9) have been prepared.

[93] HEPWORTH, M. A., R. D. PEACOCK, and P. L. ROBINSON: J. chem. Soc. **1954**, 1197.
[94] PEACOCK, R. D.: J. Chem. Soc. **1955**, 602.

The following are illustrative for metathetical reactions in bromine(III) fluoride.

$$NOSO_3F + AgBrF_4 \rightleftharpoons NOBrF_4 + AgSO_3F$$
$$KSO_3F + BrF_2SbF_6 \rightleftharpoons BrF_2SO_3F + KSbF_6$$
$$NOSO_3F + BrF_2AsF_6 \rightleftharpoons BrF_2SO_3F + NOAsF_6$$

Table 38. *Examples of the Preparation of Fluoro-complexes in Liquid Bromine(III) fluoride*

Reactants for		Reaction Product
basic solution	acid solution	
AgCl	Sb_2O_3	$AgSbF_6$
Ag	Au	$AgAuF_4$
$BaCl_2$	VCl_3	$Ba(VF_6)_2$
Ag	VCl_3	$AgVF_6$
NOCl	V_2O_5	$NOVF_6$
$CaCl_2$	Nb	$Ca(NbF_6)_2$
RbBr	Ta	$RbTaF_6$
NaCl	B_2O_3	$NaBF_4$
NOCl	GeO_2	$(NO)_2GeF_6$
NOCl	PBr_5	$NOPF_6$
KCl	$SiCl_4$	K_2SiF_6
NOCl	Au	$NOAuF_4$
AgBr	As_2O_3	$AgAsF_6$
NO_2	As_2O_3	NO_2AsF_6
KBr	$PtBr_4$	K_2PtF_6
KBr	Ru	$KRuF_6$
NO_2	SO_3	NO_2SO_3F
KCl	$Mn(IO_3)_2$	$KMnF_5$
AgCl	Cr_2O_3	$AgCrOF_4$

B. Iodine(V) fluoride

Iodine(V) fluoride is a mild fluorinating agent and a solvent which is useful for the preparation of certain fluoro-complexes[95-101] although the number of soluble compounds is much smaller than in bromine(III) fluoride.

Table 39. *Some Physical Properties of Iodine(V) fluoride*

Melting Point (°C)	+9.4
Boiling Point (°C)	+100.5
Enthalpy of Evaporation (kcal · mole^{-1})	9.8
Trouton-Constant (cal · deg^{-1} · mole^{-1})	27.3
Specific Conductivity (Ohm^{-1} · cm^{-1})	5.10^{-6}
Dielectric Constant at 35°	36.2
Viscosity at 25° (Centipoise)	2.19

[95] EMELÉUS, H. J., and G. SHARPE: J. Chem. Soc. **1949**, 2206.

[96] HARGREAVES, G. B., and R. D. PEACOCK: J. Chem. Soc. **1960**, 2373.

[97] ROGERS, M. T., and J. J. KATZ: J. Amer. Chem. Soc. **74**, 1375 (1952).

[98] ROGERS, M. T., H. B. THOMPSON, and J. L. SPEIRS: J. Amer. Chem. Soc. **76**, 4841 (1954).

[99] ROGERS, M. T., J. L. SPEIRS, H. B. THOMPSON, and M. B. PANISH: J. Amer. Chem. Soc. **76**, 4841 (1954).

[100] ROGERS, M. T., and E. E. GARRER: J. Phys. Chem. **62**, 952 (1958).

[101] WOOLF, A. A.: J. Chem. Soc. **1950**, 1053, 3678.

Iodine(V) fluoride is in the liquid state highly associated and fluorine exchange is continuously taking place in the pure liquid. The dielectric constant is reasonably high. It has a tendency to accept fluoride ions, to give IF_6^- ions, which seem to be formed also by self-ionization of the pure liquid:

$$2\,IF_5 \rightleftharpoons IF_4^+ + IF_6^-$$

Even although the model of the self-ionization is not established beyound doubt, it does provide a basis on which useful studies can be made. The compounds of composition $ISbF_{10}$ and KIF_6 may have the structures $IF_4^+SbF_6^-$ and $K^+IF_6^-$ respectively and can be made to react in liquid iodine(V) fluoride according to the equation:

$$KIF_6 + IF_4SbF_6 \rightleftharpoons KSbF_6 + 2\,IF_5$$

KIF_6 has also been obtained by the action of iodine(V) fluoride on potassium nitrate[102] or potassium iodide[96]. Fluorine exchange is also observed in solutions of hydrogen fluoride in iodine(V) fluoride[97].

In iodine(V) fluoride, iodine forms a blue solution which contains the uncoordinated I_3^+-ion[103]:

$$6\,IF_5 + 7\,I_2 \rightleftharpoons 5\,I_3^+ + 5\,IF_6^-$$

Blue solutions are also formed from I_2O_4 and IF_5. The acceptor properties of iodine(V) fluoride are also shown by the formation of an adduct with dioxane[104].

C. Arsenic(III) fluoride

Arsenic(III) fluoride is similar in its solvent properties to iodine(V) fluoride. Association in the liquid state seems to be due to fluorine bridging. The occurrence of F-exchange has been concluded to take place from the results of n.m.r. measurements[105]. A self-ionization equilibrium is assumed to be present in the pure liquid[106] due to autofluoridolysis[68, 69]

$$
\underbrace{AsF_3}_{\text{base I}} + \underbrace{AsF_3}_{\text{acid II}} \rightleftharpoons \underbrace{AsF_2^+}_{\text{acid I}} + \underbrace{AsF_4^-}_{\text{base II}}
$$

which may easily be explained by assuming a liquid structure like

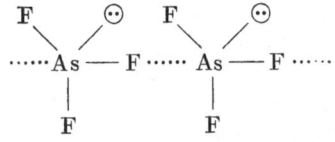

[102] AYNSLEY, E. E., R. NICHOLS, and P. L. ROBINSON: J. Chem. Soc. **1953**, 623.
[103] AYNSLEY, E. E., N. N. GREENWOOD, and D. H. W. WARMBY: J. Chem. Soc. **1963**, 5369.
[104] SCOTT, A. F., and J. F. BUNNETT: J. Amer. Chem. Soc. **64**, 2727 (1942).
[105] MUETTERTIES, E. L., and W. D. PHILLIPS: J. Amer. Chem. Soc. **79**, 3686 (1957).
[106] WOOLF, A. A., and N. N. GREENWOOD: J. Chem. Soc. **1950**, 2200.

Table 40. *Some Physical Properties of Arsenic(III) fluoride*

Melting Point (°C)	− 8.5
Boiling Point (°C)	+57.8
Dipole Moment at 20° (Debye)	2.6
Viscosity at 8° (Centipoise)	1.17
Specific Conductivity at 25° (Ohm^{-1} · cm^{-1})	$2.2 \cdot 10^{-5}$
Enthalpy of Vaporization (kcal · mole^{-1})	8.57

Arsenic(III) fluoride solvates anions, such as fluoride ions to give [AsF$_4$]$^-$ as well as some covalent compounds[107] with the exception of alkali metal fluorides. The AsF$_4^-$-ion is isoelectronic with selenium(IV) fluoride, which has a distorted trigonal bipyramidal structure. Ionic compounds are usually insoluble. Various other halides are fluorinated such as phosphorus(III) chloride, phosphorus(V) chloride, silicon(IV) chloride, thionyl chloride or tungsten hexachloride.

Neutralization type reactions, such as

$$K^+AsF_4^- + AsF_2^+SbF_6^- \rightleftharpoons KSbF_6 + 2\,AsF_3$$

have been described. The AsF$_2^+$-ion might be stabilized in solution by a bridged structure involving solvent molecules but it has not been found in a crystal so far. In solutions of antimony(V) fluoride various fluorine-bridged structures have been discussed[105], such as

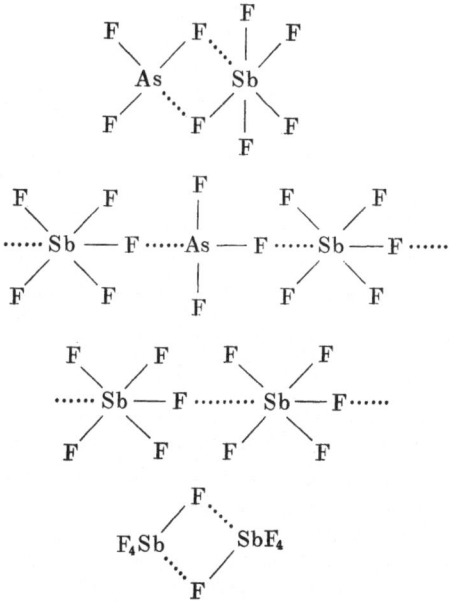

Arsenic(III) fluoride reacts with chlorine to give [AsCl$_4$][AsF$_6$][108], but small amounts of water are essential to ensure a smooth reaction[109].

[107] RUFF, O.: Ber. dtsch. chem. Ges. **37**, 4513 (1904).
[108] KOLDITZ, L.: Z. anorg. allg. Chem. **280**, 313 (1955).
[109] DESS, H. M., R. W. PARRY, and G. L. VIDALE: J. Amer. Chem. Soc. **78**, 5730 (1956).

Arsenic(III) fluoride is a good fluorinating agent for non-metal chlorides, but the reactions are sometimes incomplete. Antimony(V) chloride is converted into $SbCl_4^+F^-$ [110]; phosphorus(V) chloride gives on treatment with arsenic(III) fluoride the compound $[PCl_4][PF_6]$ and niobium(V) chloride gives $NbCl_4F$ [111].

$$[PCl_4][PCl_6] + 2\,AsF_3 \rightarrow [PCl_4][PF_6] + 2\,AsCl_3$$

3. Covalent Chlorides

Certain covalent acceptor chlorides such as molten iodine monochloride[112,113], arsenic(III)chloride[114] and antimony(III)chloride[115-117] are suitable solvents for the formation of various chloro-complexes. Hydrogen chloride[118,119] may also be included in this group of solvents, but has been discussed in chapter IV as a protonic acceptor solvent.

Self-ionization equilibria have been assumed to exist in the pure liquids. DAVIES and BAUGHAN[120] have shown that the "self-ionization" in molten antimony(III) chloride is likely to be principally due to the presence of small amounts of impurities, but the results are still at least partially in accord with the following equations, which have been regarded as due to chloride ion transfer reactions between solvent molecules[68, 69]:

$$
\begin{array}{llllll}
\text{ICl} & + & \text{ICl} & \rightleftharpoons & \text{I}^+ & + & \text{ICl}_2^- \\
\text{AsCl}_3 & + & \text{AsCl}_3 & \rightleftharpoons & \text{AsCl}_2^+ & + & \text{AsCl}_4^- \\
\text{SbCl}_3 & + & \text{SbCl}_3 & \rightleftharpoons & \text{SbCl}_2^+ & + & \text{SbCl}_4^- \\
2\,\text{HCl} & + & \text{HCl} & \rightleftharpoons & \text{H}_2\text{Cl}^+ & + & \text{HCl}_2^-
\end{array}
$$

Cl-donor	Cl-acceptor		(acid I)	(base II)
(base I)	(acid II)			

Table 41. *Some Physical Properties of Iodine Monochloride, Arsenic(III) chloride and Antimony(III) chloride*

Property	ICl	AsCl₃	SbCl₃
Melting Point (°C)	+27.2	−18.0	+73.0
Boiling Point (°C)	+100	+100	+221
Density (g · cm⁻³)	3.13 (45°)	2.16 (20°)	2.44 (78°)
Dielectric Constant	—	12.4 (21°)	33.2 (78°)
Spec. Conductivity (Ohm⁻¹ · cm⁻¹)	$4 \cdot 10^{-3}$ (35°)	$1 \cdot 10^{-7}$ (21°)	$8 \cdot 10^{-7}$ (98°)
Viscosity (Centipoise)	4.19 (28°)	1.22 (20°)	8.3 (95°)

[110] KOLDITZ, L.: Z. anorg. allg. Chem. **289**, 128 (1957).

[111] KOLDITZ, L., and G. FURCHT: Z. anorg. allg. Chem. **312**, 11 (1961).

[112] GUTMANN, V.: Research **3**, 337 (1950).

[113] GUTMANN, V.: Z. anorg. allg. Chem. **264**, 165 (1951).

[114] GUTMANN, V.: Z. anorg. allg. Chem. **266**, 331 (1951).

[115] JANDER, G., and K. H. SWART: Z. anorg. allg. Chem. **299**, 252 (1959).

[116] JANDER, G., and K. H. SWART: Z. anorg. allg. Chem. **301**, 54 (1959).

[117] JANDER, G., and K. H. SWART: Z. anorg. allg. Chem. **301**, 80 (1959).

[118] WADDINGTON, T. C.: Trans. Farad. Soc. **54**, 25 (1958).

[119] PEACH, M. E., and T. C. WADDINGTON: Chapter 3 in "Non-Aqueous Solvent Systems," Ed. T. C. WADDINGTON, Acad. Press 1965, London-New York.

[120] DAVIES, A. G., and E. C. BAUGHAN: J. Chem. Soc. **1961**, 1711.

Several review articles are available[57,121-124]. The solvents have practically no tendency to solvate metal cations[125], although they are polar molecules. Because of their acceptor properties they can easily solvate chloride ions. Complete exchange between the chloride ions of $(C_2H_5)_4NCl$ and the arsenic(III) chloride is found to occur in agreement with the existence of $AsCl_4^-$-ions[126].

Various compounds such as $K^+ICl_2^-$ are known to contain the linear $[ICl_2]^-$ ion in the crystalline state. Likewise tetrachloroarsenates may be obtained from tetraalkylammonium chloride solutions in liquid arsenic(III) chloride. The $AsCl_4^-$-ion is isoelectronic with selenium(IV) chloride, which has a distorted bipyramidal structure. The compounds $(C_2H_5)_4NCl \cdot 2AsCl_3$ and $(C_2H_5)_4NCl \cdot 3AsCl_3$ are also known[127]. Two of the arsenic(III) chloride molecules are loosely bound in the latter compound and are lost by heating at 100° at a pressure of 9 Torr. Evidence is obtained from the phase diagram for the existence of $(C_2H_5)_4NCl \cdot 5AsCl_3$[128]. With antimony(III) chloride the solvates are numerous, for example: M^ISbCl_4, $M_2^ISbCl_5$, $M^ISb_2Cl_7$, $M_3^ISb_2Cl_9$ ($M^I = Li$, K, NH_4, Rb or Cs); $M^{II}[SbCl_4]_2$, $M^{II}SbCl_5$ ($M^{II} =$ Be, Mg, Ca, Sr or Ba). A wide variety of other addition compounds with antimony(III)chloride has been reported[124].

The electrical conductivities of the pure liquid solvents are considerably increased by dissolving the respective solvates or the alkali chlorides in iodine monochloride[113,129], arsenic(III) chloride[130] or antimony(III) chloride[115]. The respective solvent anions are produced in such solutions, which have therefore basic character. The same species common to such solutions are also produced, when phosphorus(V)chloride is dissolved in the respective solvents[112,131-134] allowing phosphorus to achieve tetrahedral coordination:

$$PCl_5 + ICl \rightleftharpoons PCl_4^+ + ICl_2^- \quad \text{or}$$
$$PCl_4ICl_2 \rightleftharpoons PCl_4^+ + ICl_2^-$$
$$PCl_5 + AsCl_3 \rightleftharpoons PCl_4^+ + AsCl_4^- \quad \text{or}$$
$$(PCl_5)_2(AsCl_3)_5 \rightleftharpoons 2PCl_4^+ + 2AsCl_4^- + 3AsCl_3$$
$$PCl_5 + HCl \rightleftharpoons PCl_4^+ + HCl_2^-$$

On the other hand phosphorus(V) fluoride gives the hexafluorophosphate ion in solutions of bromine(III) fluoride. Thus phosphorus(V) chloride is a solvo-

[121] GREENWOOD, N. N.: Rev. pure appl. Chem. 1, 84 (1951).

[122] GREENWOOD, N. N. in Supplement to Mellors "Comprehensive Treatise on Inorganic and Theoretical Chemistry" II, Part I (1956).

[123] SHARPE, A. G.: Chapter 7 in "Non-Aqueous Solvent Systems," Ed. T. C. WADDINGTON, Academic Press, London-New York 1965.

[124] PAYNE, D S.: Chapter 8 in "Non-Aqueous Solvent Systems," Ed. T. C. WADDINGTON, Academic Press, London-New York 1965.

[125] HOLMES, R. R.: J. Inorg. Nucl. Chem. 12, 266 (1960).

[126] LEWIS, J., and D. B. SOWERBY: J. Chem. Soc. 1957, 336.

[127] LINDQVIST, I., and L. H. ANDERSSON: Acta Chem. Scand. 8, 128 (1954).

[128] AGERMAN, M., L. H. ANDERSSON, I. LINDQVIST, and M. ZACKRISSON: Acta Chem. Scand. 12, 477 (1958).

[129] CORNOG, J., and R. A. KARGES: J. Amer. Chem. Soc. 54, 1882 (1932).

[130] GUTMANN, V.: Mh. Chem. 85, 491 (1954).

[131] FIALKOV, Y. A., and A. A. KUZMENKO: Zh. Obshch. Khim. 19, 1645 (1949).

[132] FIALKOV, Y. A., and A. A. KUZMENKO: Zh. Obshch. Khim. 19, 812 (1949).

[133] ZELEZNY, W. F., and N. C. BAENZINGER: J. Amer. Chem. Soc. 74, 6151 (1952).

[134] GUTMANN, V.: Mh. Chem. 83, 583 (1952).

base[113,134] just as are nitrogenous bases, such as pyridine, trimethylphosphine and triethylamine[135,136] again demonstrating the acceptor properties of the solvent molecules.

Triphenylchloromethane, which is known to ionize in acceptor solvents acts also as a solvo-base by donating chloride ions to the solvent molecules:

$$(C_6H_5)_3CCl + AsCl_3 \rightleftharpoons (C_6H_5)_3C^+AsCl_4^-$$

and nitrosyl chloride behaves similarly[137]:

$$NOCl + AsCl_3 \rightleftharpoons NO^+ + AsCl_4^-$$

The high conductivities of the solutions of alkali chlorides in molten iodine monochloride would permit the assumption that the chloride ions possess abnormally high mobilities. This has been shown to occur in arsenic(III) chloride[69] and antimony(III) chloride[138] by the results of transport number measurements. A chain-conduction mechanism involving chloride ions may be in operation.

The chloride ion activity of the solutions may be represented by the pCl^--value[68, 69]:

$$p_{Cl^-} \equiv -\log a_{Cl^-}$$

Precise measurements were impossible due to failure to prepare truly reversible electrodes. Potentiometric titrations have, however, been successfully carried out in solutions of anhydrous arsenic(III) chloride employing silver-silver chloride electrodes[139].

Acidic behaviour has been found for various acceptor compounds, but they are unsolvated and poor conductors in the solutions. The following were found to behave as weak acids: aluminium chloride, ferric chloride, stannic chloride, titanium(IV) chloride[113,114], vanadium(IV) chloride[140], tellurium(IV) chloride[141], and antimony(V) chloride[113,114].

Neutralization reactions lead to the formation of chloro-complexes:

$$SbCl_5 + ICl_2^- \rightleftharpoons SbCl_6^- + ICl$$
$$SbCl_5 + AsCl_4^- \rightleftharpoons SbCl_6^- + AsCl_3$$
$$SbCl_5 + SbCl_4^- \rightleftharpoons SbCl_6^- + SbCl_3$$

Conductometric titrations show that the formation of hexachlorostannates, hexachlorotellurates and of hexachlorotitanates occurs in two steps in molten iodine monochloride[113] and in liquid arsenic(III) chloride[114,141]. At a 1 : 1-molar ratio of the basic chloride and the acceptor chloride the solution reaches a maximum in conductivity and with further addition of the chloride ion-donor insoluble hexachlorometallate is formed, allowing the conductivity to pass through a minimum. The formation of an "acidic-salt" has been postulated, but conclusive evidence is not available.

[135] HOLMES, R. R., and E. F. BERTAUT: J. Amer. Chem. Soc. **80**, 2980, 2983 (1958).
[136] HOLMES, R. R.: J. Amer. Chem. Soc. **82**, 5285 (1960).
[137] GUTMANN, V.: J. Med. Fac. Baghdad **16**, 132 (1953).
[138] FRYCZ, K., and S. TOLLOCZKO: Chem. Zbl. **84**, I, 91 (1913).
[139] ANDERSSON, L. H., and I. LINDQVIST: Acta Chem. Scand. **9**, 79 (1953).
[140] GUTMANN, V.: Mh. Chem. **85**, 286 (1954).
[141] GUTMANN, V.: Mh. Chem. **83**, 159 (1952).

$$[R_4N]^+[AsCl_4]^- + TiCl_4 \rightleftharpoons [R_4N][AsCl_2][TiCl_6]$$
$$[R_4N][AsCl_2][TiCl_6] + [R_4N]^+[AsCl_4]^- \rightleftharpoons [R_4N]_2[TiCl_6] + 2\,AsCl_3$$

Redox reactions involving free chlorine in iodine monochloride lead to the formation of iodine(III) chloride that appears to act as a weak base in this medium:

$$ICl + Cl_2 \rightleftharpoons ICl_3$$
$$ICl_3 + ICl \rightleftharpoons ICl_2^+ + ICl_2^-$$

When chlorine is passed through a solution of arsenic(III) chloride, which contains an acceptor chloride, complex compounds of the unknown arsenic(V) chloride are formed, which presumably contain $[AsCl_4]^+$ units[142,143]

$$AsCl_3 + Cl_2 + SbCl_5 \rightleftharpoons [AsCl_4][SbCl_6]$$
$$AsCl_3 + Cl_2 + FeCl_3 \rightleftharpoons [AsCl_4][FeCl_4]$$
$$AsCl_3 + Cl_2 + AlCl_3 \rightleftharpoons [AsCl_4][AlCl_4]$$

The $AsCl_4^+$ unit has also been assumed to exist in the compound $[AsCl_4][AsF_6]$[144]. This interesting compound which is somewhat soluble in liquid arsenic(III) chloride and very soluble in liquid arsenic(III) fluoride is formed when chlorine is passed through liquid arsenic(III) fluoride at $0°$:

$$2\,AsF_3 + 2\,Cl_2 \rightleftharpoons [AsCl_4][AsF_6]$$

The compound $[PCl_4][PCl_5Br]$ is formed, when phosphorus(III) chloride is reacted with bromine in arsenic(III) chloride[145]. KOLDITZ[146] has also used arsenic(III)-chloride as a solvent to convert $[PCl_4][PF_6]$ into a mixture of ionic and molecular PCl_4F. It has further been shown that $ZrCl_4 \cdot PCl_5$ is converted into $ZrCl_2F_2$ by treatment with arsenic(III) fluoride in arsenic(III) chloride[147].

Due to the high reactivity of molten iodine monochloride various elements are easily chlorinated[148], although the chlorinating properties of iodine monochloride are weaker than the fluorinating power of bromine(III) fluoride.

$$K + 2\,ICl \rightarrow K^+ICl_2^- + \tfrac{1}{2}I_2$$
$$P + 6\,ICl \rightarrow PCl_4^+ICl_2^- + {}^5/_2 I_2$$

Likewise arsenic(III)chloride reacts with sodium:

$$Na + AsCl_3 \rightarrow As + 3\,NaCl$$

Vanadium(III) chloride may be obtained in small quantities from finely devided vanadium and iodine(I) chloride[149] since vanadium(IV) chloride is reduced by iodine:

$$V + 3\,ICl \rightarrow VCl_3 + {}^3/_2 I_2$$
$$VCl_4 + {}^1/_2 I_2 \rightleftharpoons VCl_3 + ICl$$

[142] GUTMANN, V.: Mh. Chem. **82**, 473 (1950).
[143] KOLDITZ, L., and W. SCHMIDT: Z. anorg. allg. Chem. **296**, 188 (1958).
[144] KOLDITZ, L.: Z. anorg. allg. Chem. **280**, 313 (1955).
[145] KOLDITZ, L., and A. FELTZ: Z. anorg. allg. Chem. **293**, 286 (1957).
[146] KOLDITZ, L.: Z. anorg. allg. Chem. **286**, 307 (1956).
[147] KOLDITZ, L., and P. DEGENKOLB: Z. Chem. (Leipzig) **6**, 347 (1966).
[148] GUTMANN, V.: Z. anorg. allg. Chem. **264**, 169 (1951).
[149] GUTMANN, V.: Mh. Chem. **81**, 1155 (1950).

Several metal oxides are also converted into the respective chlorides[150]. Examples of other reactions in arsenic(III) chloride are represented by the following equations:

$$PI_3 + AsCl_3 \rightleftharpoons AsI_3 + PCl_3$$
$$3\,R_4NI + 4\,AsCl_3 \rightleftharpoons 3\,[R_4N]^+ + 3\,[AsCl_4]^- + AsI_3$$
$$n\,As_2O_3 + n\,AsCl_3 \rightarrow [AsOCl)_{3n}$$

The latter compound probably consists of chains[151]:

$$
\begin{array}{c}
Cl \\
| \\
-As-O-As-O- \\
| \\
Cl
\end{array}
$$

In antimony(III) chloride the following reactions are known to occur:

$$3\,KBr + SbCl_3 \rightleftharpoons 3\,KCl + SbBr_3$$
$$Sb_2S_3 + SbCl_3 \rightleftharpoons 3\,SbSCl$$

K_2TiCl_6 has been readily prepared in antimony(III) chloride:

$$2\,K^+SbCl_4^- + TiCl_4 \rightleftharpoons K_2TiCl_6 + 2\,SbCl_3$$

but K_3TiCl_6 is not obtained in this solvent from potassium chloride and titanium(III) chloride because the solvent is a better chloride ion acceptor than titanium(III) chloride and the only salt prepared was $(C_2H_5)_4N^+SbCl_4^-$ [152].

4. Covalent Bromides

Just as certain chlorides are found to behave as useful solvents for the formation of chloro-complexes, bromo-complexes are formed in solutions of the corresponding bromides. The solvent behaviour of (for example) molten iodine monobromide[153], molten arsenic(III) bromide[154-156], aluminium(III) bromide[157] or molten mercury(II) bromide[158-163] has been described; the chemistry in such solutions has been briefly reviewed[73,123,124]. Hydrogen bromide is very similar, but has been discussed separately in this presentation as a protonic acceptor solvent (chapter IV).

[150] KANE, R.: Phil. Mag. (3) **10**, 430 (1837), J. prakt. Chem. **11**, 250 (1837).
[151] THILO, E., and P. FLÖGEL: Angew. Chem. **69**, 754 (1957).
[152] FOWLES, G. W. A., and B. J. RUSS: J. Chem. Soc. (A) **1967**, 517.
[153] GUTMANN, V.: Mh. Chem. **82**, 156 (1951).
[154] JANDER, G., and K. GÜNTHER: Z. anorg. allg. Chem. **297**, 81 (1958).
[155] JANDER, G., and K. GÜNTHER: Z. anorg. allg. Chem. **298**, 241 (1959).
[156] JANDER, G., and K. GÜNTHER: Z. anorg. allg. Chem. **302**, 154 (1959).
[157] JANDER, G., and W. ZSCHAAGE: Z. anorg. allg. Chem. **272**, 53 (1953).
[158] JANDER, G., and K. BRODERSEN: Z. anorg. allg. Chem. **261**, 261 (1950).
[159] JANDER, G., and K. BRODERSEN: Z. anorg. allg. Chem. **262**, 33 (1950).
[160] JANDER, G., and K. BRODERSEN: Z. anorg. allg. Chem. **264**, 57 (1951).
[161] JANDER, G., and K. BRODERSEN: Z. analyt. Chem. **133**, 146 (1951).
[162] JANDER, G., and K. BRODERSEN: Z. anorg. allg. Chem. **264**, 76 (1951).
[163] JANDER, G., and K. BRODERSEN: Z. anorg. allg. Chem. **264**, 92 (1951).

Table 42. *Some Physical Properties of* IBr, AsBr$_3$, AlBr$_3$ *and* HgBr$_2$

Property	IBr	AsBr$_3$	AlBr$_3$	HgBr$_2$
Melting Point (°C)	41	35	97	238
Boiling Point (°C)	46	220	268	320
Density (g · cm^{-3})		3.33 (50°)		5.109
Dielectric Constant		8.8 (35°)	2.9 (265°)	9.8 (238°)
Specific Conductivity (Ohm^{-1} · cm^{-1})	3 · 10^{-4}	1 · 10^{-7}	9 · 10^{-8}	10^{-8}
Heat of Vaporization (kcal · mole^{-1})			5.62	14.1
Viscosity (Centipoise)		5.41 (35°)		3.32

The self-ionization equilibria which have been assumed to exist in the pure liquids have been considered as bromide ion transfer reactions[68,69]

$$
\begin{array}{ccccccc}
 & \overset{\displaystyle Br^-}{\overbrace{}\big\downarrow} & & & & & \\
IBr & + & IBr & \rightleftharpoons & I^+ & + & IBr_2^- \\
AsBr_3 & + & AsBr_3 & \rightleftharpoons & AsBr_2^+ & + & AsBr_4^- \\
AlBr_3 & + & AlBr_3 & \rightleftharpoons & AlBr_2^+ & + & AlBr_4^- \\
HgBr_2 & + & HgBr_2 & \rightleftharpoons & HgBr^+ & + & HgBr_3^-
\end{array}
$$

bromide ion-donor (base) bromide ion-acceptor (acid)

The IBr$_2^-$-ion is linear like the ICl$_2^-$-ion; the AsBr$_4^-$-ion may be expected to have a distorted bipyramidal structure analogous to that of the tetrachloro-arsenate(III)-ion. Although the molecules are polar, they do not interact appreciably with metal cations and solvation is known to occur in the first place with donor ions or donor molecules, such as bromide ions and certain N-bases. The compounds R$_4$NBr · 3AsBr$_3$, PyHAsBr$_4$ and (CH$_3$)$_3$NHBr · AsBr$_3$ which have been prepared[154] are likely to contain solvated bromide ions.

Iodine bromide and arsenic(III) bromide resemble closely as solvents iodine monochloride and arsenic(III) chloride respectively, and stable acids have not been isolated in the bromo-systems. Molten mercury(II) bromide is an excellent solvent for various classes of compounds. Alkali metal bromides form solvated anionic species which appear to contain the ions [IBr$_2$]$^-$, [AsBr$_4$]$^-$ and [HgBr$_3$]$^-$ respectively and phosphorus(V) bromide gives PBr$_4^+$IBr$_2^-$ in iodine bromide where it behaves as a base. Other basic substances are nitrogenous bases such as pyridine.

In aluminium(III) bromide alkali halides are usually soluble with the formation of tetrabromoaluminates just as well as in molten mercuric bromide, when tri-bromomercurates may be isolated from the solutions.

All soluble bromides of the alkali metals and alkaline earth metals behave as solvo-bases in the respective covalent bromide solvent systems under consideration. While in iodine bromide and in aluminium bromide only few and very weak solvo-acids have been found, various acceptor bromides react readily with bromides in solutions of arsenic(III) bromide and mercury(II) bromide.

The reaction of gold(III) bromide with tetraethylammonium bromide in arsenic(III) bromide has been followed conductometrically and tetraethylammonium tetrabromoaurate(III) appears to be formed. Many other complex bromides have been isolated from the solutions, including $(C_2H_5)_3NH^+AlBr_4^-$, $(C_2H_5)_2NH_2^+AlBr_4^-$, $Pb^{++}[AlBr_4^-]_2$ (which is scarcely soluble in molten arsenic(III) chloride), $K^+AlBr_4^-$, $Cu^+AlBr_4^-$, $Ag^+AlBr_4^-$, $Cs^+GaBr_4^-$ and $K^+GaBr_4^-$.

Mercuric bromide dissolves also most of the mercuric salts, such as the perchlorate, sulphate, nitrate and phosphate which behave as solvo-acids in this solvent, e.g.

$$Hg(ClO_4)_2 + HgBr_2 \rightleftharpoons 2\,HgBr^+ + 2\,ClO_4^-$$
$$HgSO_4 + HgBr_2 \rightleftharpoons 2\,HgBr^+ + SO_4^{--}$$

Bromides of the electropositive metals including Tl^+ are typical solvo-bases in molten mercuric bromide. Mercuric oxide also behaves as a solvo-base, since it produces bromide ions by reaction with solvent molecules, e.g.:

$$HgO + 2\,HgBr_2 \rightleftharpoons Hg_2OBr^+ + HgBr_3^-$$

Ionic reactions between solvo-acids and solvo-bases may lead to insoluble products such as thallium sulphate formed from mercury(II) sulphate and thallium bromide in molten mercury(II) bromide. Similarly anhydrous copper(II) sulphate can be prepared by using a copper(II) halide. Perchlorates, nitrates and phosphates of many other elements can be prepared in a similar manner. By allowing mercury(II) oxide to react with the sulphate in mercury(II) bromide solution a red, insoluble product of composition $(HgO)_2HgSO_4$ is formed. Analogous compounds are formed from the sulphide, selenide and telluride of mercury in molten mercuric bromide.

Many conductometric titrations have been carried out in solutions of molten mercury(II) bromide[160]. For potentiometric work a gold electrode has been found to give nearly reproducible potential values[161]. It has been suggested that conductometric and potentiometric methods could be used for the estimation of certain electrolytes in molten mercury(II) bromide[161]. For example mercury(II) oxide may be readily titrated with mercury(II) perchlorate in the bromide melt.

Iodine bromide converts various metals into the bromides[164], but its reactivity is smaller than that of iodine monochloride. Arsenic(III) bromide reacts with the oxides of lead, gold or copper to give the respective bromides[159]:

$$3\,CuO + 2\,AsBr_3 = As_2O_3 + 3\,CuBr_2$$
$$CuBr_2 = CuBr + {}^1/_2\,Br_2$$
$$Au_2O_3 + 2\,AsBr_3 = 2\,AuBr_3 + As_2O_3$$
$$3\,PbO + 2\,AsBr_3 = 3\,PbBr_2 + As_2O_3$$

Arsenic(III) oxide and arsenic(III) sulphide are converted into the oxy- and thiobromides respectively:

$$As_2O_3 + AsBr_3 = 3\,AsOBr$$
$$As_2S_3 + AsBr_3 = 3\,AsSBr$$

No reactions were observed with the oxides of aluminium, titanium, chromium(III) and iron(III), but certain carbonates and sulphates were found to react:

$$3\,CdCO_3 + 2\,AsBr_3 = 3\,CdBr_2 + As_2O_3 + 3\,CO_2$$
$$3\,K_2SO_4 + 2\,SbBr_3 = 6\,KBr + Sb_2(SO_4)_3$$

[164] GUTMANN, V.: Mh. Chem. **82**, 169 (1951).

Aluminium(III) bromide converts sodium, calcium, indium and thallium into the respective bromides:

$$3\,Na + AlBr_3 = 3\,NaBr + Al$$

The sulphides of silver, mercury, lead and antimony are appreciably soluble in molten aluminium bromide[157]. Carbonates are converted into the bromides with simultaneous liberation of carbon dioxide. Thionyl chloride yields thionyl bromide:

$$3\,SOCl_2 + 2\,AlBr_3 \rightleftharpoons 3\,SOBr_2 + 2\,AlCl_3$$

and potassium fluoride the sparingly soluble aluminium fluoride[157]:

$$3\,KF + AlBr_3 = 3\,KBr + AlF_3$$

Mercury(II) bromide acts as a good dehydrating agent for bromides containing water of crystallization, which is volatile at temperatures corresponding with the liquid range of mercury(II) bromide. Thus anhydrous salts may be prepared from the hexahydrate of mercury(II) perchlorate and from the dihydrate of mercury(II) nitrate.

5. Molten Iodine

In the solid state the interatomic distance in the iodine molecule (2.68 Å) is not a great deal shorter than the distance between iodine atoms of different molecules (3.56 Å). The intermolecular attraction seems also to persist in the liquid state as suggested by the high heat of vaporization at the boiling point.

Table 43. *Some Physical Properties of Iodine*

Melting Point (°C)	+114.0
Boiling Point (°C)	+183.0
Density at 133° $(g \cdot cm^{-3})$	3.92
Dielectric Constant at 118°	11.1
Specific Conductivity at 114° $(Ohm^{-1} \cdot cm^{-1})$ ...	$5 \cdot 10^{-5}$
Viscosity at 116° (Centipoise)	1.98
Enthalpy of Vaporization $(kcal \cdot mole^{-1})$	10.6

The conductivity of molten iodine has at least in part been ascribed to the self-ionization equilibrium[165]:

$$2\,I_2 \rightleftharpoons I^+ + I_3^-$$

This equilibrium may be regarded as due to an iodide ion transfer[68] but it clearly involves a redox reaction, since the oxidation numbers of iodine are changed in the course of this reaction. Iodine is an acceptor molecule to give the tri-iodide ion or other polyhalide ions with iodine.

$$I_2 + I^- \rightleftharpoons I_3^-$$

It acts as an acceptor towards alcohols, ethers, esters, ketones and nitrogenous bases as well as towards π-donors such as benzene as can be seen from the charac-

[165] JANDER, G., and K. H. BANDLOW: Z. physik. Chem. (A) **191**, 321 (1942).

teristic charge-transfer-bands of the adducts in solution[166],[167]. Adducts are known to be formed from iodine and various molecules, such as trimethylphosphate, dimethylsulphoxide or many amines where the heats of formation are between 2 and 10 kcal/mole[168].

The alkali iodides are soluble in molten iodine to give conducting solutions typical for weak 1 : 1 electrolytes. Alkali iodides have been regarded as solvo-bases and iodine monohalides as solvo-acids. A conductometric titration of potassium iodide with iodine monobromide in molten iodine shows a significant break at a molar ratio KI : IBr = 1 : 1, indicating the reaction:

$$KI_3 + IBr \rightleftharpoons KBr + 2I_2$$

This process has been followed potentiometrically, although iodine monohalides are only weak electrolytes in iodine, for which an ionic product

$$K = c_{I^+} \cdot c_{I_3^-} \sim 10^{-42}$$

has been estimated[165].

Solvolytic reactions and complex formation have also been reported in this solvent[165]:

$$KCN + I_2 \rightarrow KI + ICN$$
$$HgI_2 + 2KI \rightarrow K_2HgI_4$$

[166] BENESI, H. A., and J. H. HILDEBRAND: J. Amer. Chem. Soc. 71, 2703 (1949).

[167] BRIEGLEB, G.: „Elektronen-Donator-Acceptor-Komplexe", Springer-Verlag, Berlin-Göttingen-Heidelberg 1961, p. 28 ff.

[168] TSUBOMURA, H.: J. Amer. Chem. Soc. 82, 40 (1960).

Chapter VI

Oxyhalide Solvents

Oxyhalide-solvents can act as halide ion-donors in a similar manner to the acceptor halides discussed in the previous chapter. On the other hand they differ from the covalent halides by the donor properties at the O-atoms and coordination chemistry in their solutions will depend on the donor properties as expressed by the donor numbers[1, 2].

Table 44

Donor Numbers and Dielectric Constants of Oxyhalide Solvents

Solvent	DN_{SbCl_5}	Dielectric Constant
$COCl_2$	0	4.7 (0°)
$NOCl$	0	19.7 (−10°)
$SOCl_2$	0.1	9.0 (22°)
SO_2Cl_2	0.4	9.1 (20°)
CH_3COCl	1.0	15.0 (20°)
C_6H_5COF	2	22.7 (20°)
C_6H_5COCl	2.3	15.0 (20°)
C_6H_5COBr	∼2.5	21.0 (20°)
$POCl_3$	11.7	13.9 (22°)
$SeOCl_2$	12.2	46.2 (20°)
$C_6H_5POF_2$	16.4	28.6 (20°)
$C_6H_5POCl_2$	18.5	26.0 (20°)
$(C_6H_5)_2POCl$	19.2	

According to the different donor properties found in this group of solvents a distinction may be made between (a) oxyhalides with low donor numbers including carbonyl chloride, nitrosyl chloride, thionyl chloride, sulphuryl chloride, acetyl and benzoyl halides and (b) oxyhalides with medium donor numbers, namely phosphorus oxychloride, selenium oxychloride and phenyl phosphonic halides, the latter having donor properties approaching those of water or of the ethers.

Their properties as donor molecules are well established but it is difficult to decide, whether and to what extent impurities are responsible for the availability of halide ions in the pure solvents. It has recently been stated, that the solvent-

[1] GUTMANN, V., and E. WYCHERA: Inorg. Nucl. Chem. Letters **2**, 257 (1966).
[2] GUTMANN, V., and E. WYCHERA: Rev. Chim. Min. **3**, 941 (1966).

system concept based on the mode of self-ionization applied to oxyhalide systems has been overemphasized[3-5], but it definitely served of tremendous value in the exploration of reactions in a number of non-aqueous solvents[6-8].

1. Oxyhalides with Low Donor Numbers

Oxyhalide solvents with weak donor properties at the O-atom will be similar in solvent properties to acceptor halides discussed in the previous chapter because coordination of solvent molecules to an acceptor compound will hardly be any competition for the formation of a halide complex by accepting such ions, as might be provided in the solutions.

Carbonyl chloride is of historical interest since it was the first oxyhalide known to behave as an ionizing solvent[9-16]. Complex formation between aluminium chloride and calcium chloride was explained by assuming a self-ionization equilibrium of the solvent molecules.

Thionyl[17,18] and sulphuryl chlorides[19,20] have exceedingly low donor properties, but strong chloride ion-donor properties and these are useful media for the formation of chloro-complexes. Liquid nitrosyl chloride does not show any O-donor properties, but it is easily ionized[21,22]:

$$NOCl \rightleftharpoons NO^+ + Cl^-$$

The chloride ions provided by this self-ionization may be coordinated by acceptor compounds

$$NOCl + SbCl_5 \rightleftharpoons [NO]^+[SbCl_6]^-$$

Thus "nitrosyl chloride solvates" are essentially ionic in nature.

[3] MEEK, W. D., and R. S. DRAGO: J. Amer. Chem. Soc. **83**, 4322 (1961).

[4] DRAGO, R. S., and K. F. PURCELL: Progr. Inorg. Chem. **6**, 271 (1964).

[5] DRAGO, R. S., and K. F. PURCELL: Chapter 5 in "Non-Aqueous Solvent Systems," Ed. T. C. WADDINGTON, Academic Press London-New York, 1965.

[6] See L. F. AUDRIETH, and J. KLEINBERG: "Non-Aqueous Solvents," J. Wiley and Sons, New York, 1953.

[7] PAYNE, D. S.: Chapter 8 in "Non-Aqueous Systems," Ed. T. C. WADDINGTON, Academic Press, London-New York, 1965.

[8] GUTMANN, V.: "Halogen Chemistry," Ed. V. GUTMANN, Vol. II, 399 ff., Academic Press, London-New York, 1967.

[9] GERMANN, A. F. O.: J. Phys. Chem. **28**, 879 (1924).

[10] GERMANN, A. F. O.: J. Phys. Chem. **29**, 1148 (1925).

[11] GERMANN, A. F. O.: J. Amer. Chem. Soc. **47**, 2469 (1923).

[12] GERMANN, A. F. O.: Science **61**, 70 (1925).

[13] GERMANN, A. F. O., and K. GAGOS: J. Phys. Chem. **28**, 965 (1924).

[14] GERMANN, A. F. O., and G. H. McINTYRE: J. Phys. Chem. **29**, 139 (1925).

[15] GERMANN, A. F. O., and Q. W. TAYLOR: J. Amer. Chem. Soc. **48**, 1154 (1926).

[16] GERMANN, A. F. O., and C. R. TIMPANY: J. Amer. Chem. Soc. **47**, 2275 (1925).

[17] SPANDAU, H., and E. BRUNNECK: Z. anorg. allg. Chem. **270**, 201 (1952).

[18] SPANDAU, H., and E. BRUNNECK: Z. anorg. allg. Chem. **278**, 197 (1955).

[19] GUTMANN, V.: Mh. Chem. **85**, 393 (1954).

[20] GUTMANN, V.: Mh. Chem. **85**, 404 (1954).

[21] BURG, A. B., and G. W. CAMPBELL: J. Amer. Chem. Soc. **70**, 1964 (1948).

[22] BURG, A. B., and D. E. McKENZIE: J. Amer. Chem. Soc. **74**, 3143 (1952).

Table 45. *Some Physical Properties of Certain Oxyhalides*

Property	$COCl_2$	NOCl	$SOCl_2$	SO_2Cl_2
Melting Point (°C)	−128	−61	−104	−54
Boiling Point (°C)	+8	−5	+76	+70
Dielectric Constant	4.7 (0°)	19.7 (−10°)	9.0 (22°)	9.1 (20°)
Donor Number (DN_{SbCl_5})	0	0	0.1	0.4
Spec. Cond. (Ohm$^{-1} \cdot$ cm^{-1}) ..	10^{-9} (0°)	10^{-7} (−10°)	10^{-9} (22°)	10^{-8} (20°)

Acetyl chloride[23, 24], benzoyl fluoride[25, 26], benzoyl chloride[27-29] and benzoyl bromide[30-33] may also be used as solvents. Ionic compounds are known to be formed from acceptor compounds and such acid halides.

Table 46. *Some Physical Properties of Oxyhalides of Organic Acids*

Property	CH_3COCl	C_6H_5COF	C_6H_5COCl	C_6H_5COBr
Melting Point (°C)	−112	−28.5	−0.6	+8.1
Boiling Point (°C)	+60	+156	+197	+215
Density (g $\cdot$ cm^{-3})		1.149		1.546 (20°)
Dielectric Constant at 20°	15.0	22.7	23.0	21.0
Donor Number (DN_{SbCl_5})	2.0	2.0	2.3	
Spec. Cond. at 20° (Ohm^{-1}cm^{-1}) .	$\sim 10^{-7}$	$\sim 10^{-8}$	$\sim 10^{-8}$	$\sim 10^{-8}$

Nitrosyl chloride is associated in the liquid state. The nitrogen-chlorine bond is abnormally long (1.95 Å)[34] and the [NO]$^+$ ion is easily formed. This is assumed to be solvated by means of chlorine bridges[21, 22] and resonance between the following structures has been suggested:

$$[O=N-Cl-N=O]^+ \leftrightarrow [O=N-Cl]\overset{+}{N}\equiv O \leftrightarrow O\equiv \overset{+}{N}[Cl-N=O]$$

Alkali metal halides are practically unsolvated and hence show low solubilities in liquid nitrosyl chloride, while compounds with acceptors are usually soluble and behave as strong electrolytes in the solutions.

A strongly polarized $X=O$ bond in the oxyhalide with strong donor properties will weaken the $X-Cl$ bond, so that the possibility of the formation of halide bridges may compete with O-coordination. The adducts with nitrosyl chloride can be regarded as ionic compounds containing the nitrosonium ion [NO]$^+$ and the

[23] PAUL, R. C., D. SINGH, and S. S. SANDHU: J. Chem. Soc. **1959**, 315.
[24] SINGH, B., R. C. PAUL, and S. S. SANDHU: J. Chem. Soc. **1959**, 326.
[25] JANDER, G., and L. SCHWIEGK: Z. anorg. allg. Chem. **310**, 1 (1961).
[26] JANDER, G., and L. SCHWIEGK: Z. anorg. allg. Chem. **310**, 12 (1961).
[27] GUTMANN, V., and H. TANNENBERGER: Mh. Chem. **88**, 216 (1957).
[28] GUTMANN, V., and H. TANNENBERGER: Mh. Chem. **88**, 292 (1957).
[29] GUTMANN, V., and G. HAMPEL: Mh. Chem. **92**, 1048 (1961).
[30] GUTMANN, V., and K. UTVARY: Mh. Chem. **89**, 186 (1958).
[31] GUTMANN, V., and K. UTVARY: Mh. Chem. **89**, 731 (1958).
[32] GUTMANN, V., and K. UTVARY: Mh. Chem. **90**, 751 (1959).
[33] UTVARY, K., and V. GUTMANN: Mh. Chem. **90**, 710 (1959).
[34] GERDING, H., J. A. KONINGSTEIN, and E. R. VAN DER WORM: Spectrochim. Acta **16**, 881 (1960).

chlorometallate anions. The adduct $NOCl \cdot AlCl_3$ shows Raman lines corresponding to $[NO]^+$ and $[AlCl_4]^-$-ions, but transfer of the chloride ion seems to be incomplete[35]:

$$
\begin{array}{c}
Cl \\
| \quad - \qquad + \\
Cl-Al-Cl \cdots N = O \\
| \\
Cl
\end{array}
$$

The infrared frequency characteristic for the NO-group varies in different solvates thus indicating differences in the nature of chemical bonding, which may be accounted for by different amounts of chlorine bridging[36]. The conductivities of such compounds in liquid nitrosyl chloride are high and the solvent ions were shown to have high transport numbers, and thus abnormally high mobilities[22], which may be due to the readiness of chloride ion transfer reactions[37].

Chloride coordination also seems to be present in the adducts of aluminium chloride and gallium(III) chloride[38] with acetyl chloride and in compounds of aluminium chloride or titanium(IV) chloride with mesitoyl chloride[39, 40]. The corresponding compounds of acetyl chloride may also be considered to be ionic in character at least in solutions of liquid sulphur dioxide[41-43], while those of benzoyl chloride are represented by coordination through the oxygen atom of the solvent[44], as has been shown for $Cl_5SbOCClC_6H_5$ both by i.r.-measurements[45] and an X-ray analysis[46].

Analogous structures are assigned to benzoyl bromide solvates such as $AlBr_3 \cdot C_6H_5COBr$ as is indicated both by the comparison of the heats of solvation of bromides with keto compounds and benzoyl bromide and by the results of ultra-violet spectroscopy[47].

Few solvates have been described with benzoyl fluoride, which seems to be a poor solvent for ionic fluorides. The compounds with antimony(V) chloride and boron(III) fluoride have been represented by ionic structures[25], although no evidence is available for the solid states.

The self-ionization equilibria of the pure solvents may be described as involving transfer of chloride ions[48, 49].

[35] GERDING, H., and H. HOUTGRAAF: Rec. Trav. Chim. Pays-Bas **72**, 21 (1953).

[36] WADDINGTON, T. C., and F. KLANBERG: Z. anorg. allg. Chem. **304**, 186 (1960).

[37] GUTMANN, V.: XVII. Int. Congr. Pure and Applied Chem., Vol. I, 95, Butterworth London 1959.

[38] GREENWOOD, N. N., and K. WADE: J. Chem. Soc. **1956**, 1527.

[39] SUSZ, B. P., and J. J. WUHRMANN: Helv. Chim. Acta **40**, 971 (1957).

[40] CASSIMATIS, D., and B. P. SUSZ: Helv. Chim. Acta **44**, 943 (1961).

[41] SEEL, F.: Z. anorg. allg. Chem. **250**, 331 (1943).

[42] SEEL, F.: Z. anorg. allg. Chem. **252**, 24 (1943).

[43] SEEL, F., and H. BAUER: Z. Naturf. **2 b**, 397 (1947).

[44] RASMUSSEN, S. E., and N. C. BROCH: Chem. Comm. **1965**, 289.

[45] OLAH, G. A., S. J. KUHN, W. S. TOLGYESY, and E. B. BAKER: J. Amer. Chem. Soc. **84**, 2733 (1962).

[46] WEISS, R., and B. CHEVRIER: Chem. Comm. **1967**, 145.

[47] LEBEDEV, N. N.: J. Phys. Chem. USSR **22**, 1505 (1948). Chem. Zbl. Erg.Bd. I, **1948**, 381.

[48] GUTMANN, V., and I. LINDQVIST: Z. phys. Chem. **203**, 250 (1954).

[49] GUTMANN, V.: Svensk. Kem. Tidskr. **68**, 1 (1956).

The equilibria are usually represented as:

$$NOCl \rightleftharpoons NO^+ + Cl^-$$
$$SOCl_2 \rightleftharpoons SOCl^+ + Cl^-$$
$$SO_2Cl_2 \rightleftharpoons SO_2Cl^+ + Cl^-$$
$$RCOX \rightleftharpoons RCO^+ + X^-$$

Table 47. *Solvation Numbers in Crystalline Solvates Formed from 1 Molecule of Acceptor Halide with Oxyhalide Solvents*[8]

Solute \ Solvent	COCl₂	SOCl₂	NOCl	CH₃COCl	C₆H₅COCl
CuCl₂			1		
ZnCl₂			1		
HgCl₂			1		
PdCl₂			2		
PtCl₂			1		
MnCl₂			1		
BCl₃		1	1	1	
BF₃			1		
AlCl₃	1, 3, 5	0.5	1, 2	1	1
GaCl₃			1	1	1
InCl₃			1		
TlCl₃			1		
AsCl₃			1, 2		
SbCl₃			1, 2		
BiCl₃			1		
FeCl₃			1, 2	1	1
AuCl₃			1		
TiCl₄		1	2	1, 2	1
ZrCl₄		1	2		
ThCl₄			2		
SiCl₄			2*		
SnCl₄		1	2	2	
PbCl₄			2		
PtCl₄			2		
VCl₄			1		
UO₂Cl₂			1		
SbCl₅		1	1	1	1
SbF₅			1		

* unstable at room temperature

Considerable further contributions to the understanding of the chemistry of nitrosyl chloride solutions have been provided by the results of tracer studies[50-52] using ³⁶Cl. The exchange was studied between nitrosyl chloride and the dissolved chlorides of aluminium(III), gallium(III), indium(III), thallium(III), iron(III) and antimony(V). Rapid exchange was found and it was concluded that the chlorine

[50] LEWIS, J., and R. G. WILKINS: J. Chem. Soc. **1955**, 56.
[51] LEWIS, J., and R. G. WILKINS: J. Chem. Soc. **1957**, 336, 1617.
[52] LEWIS, J., and R. G. WILKINS: Chem. Soc. Spec. Publ. No. 10, p. 123 (1957).

atoms are all equivalent. Even chlorides of low solubility, such as zinc, cadmium and mercury(II) chlorides show rapid exchange with the solvent molecules. On the other hand no exchange was observed with alkali metal chlorides, which do not form any complex compounds with the solvent. Silver chloride was found to exchange with liquid nitrosyl chloride only in the presence of light, this probably being due to a photochemical decomposition.

Oxyhalides solvents of low donor numbers are good media for the preparation of halide complexes because coordination of acceptor compounds by solvent molecules is not taking place to any significant degree.

The formation of halide complexes has been investigated by means of preparative, conductometric, potentiometric and spectrophotometric techniques. The results are summarized in Table 48.

Table 48. *Coordination Forms of Certain Acceptor Chlorides in Weak Oxyhalide Donor Solvents in the Presence of Tetraalkylammonium Chlorides as Chloride Ion Donors*

Solvent Chloride	$SOCl_2$	SO_2Cl_2	NOCl	CH_3COCl	C_6H_5COCl
$[PCl_6]^-$	+	+			
$[SbCl_6]^-$	+	+	+	+	+
$[NbCl_6]^-$					+
$[TaCl_6]^-$					+
$[TiCl_5]^-_{sv}$	+	+		+	+
$[TiCl_6]^{2-}$	+	+	+	+	+
$[SnCl_5]^-$	+	+		+	+
$[SnCl_6]^{2-}$	+	+	+	+	+
$[ZrCl_5]^-_{sv}$	+				+
$[ZrCl_6]^{2-}$	+			+	+
$[TeCl_5]^-_{sv}$					+
$[TeCl_6]^{2-}$				+	+
$[BCl_4]^-$			+		
$[AlCl_4]^-$	+	+			+
$[FeCl_4]^-$				+	+
$[HgCl_3]^-$					+
$[HgCl_4]^{2-}$					+

The + sign in Table 48 indicates that the corresponding equilibrium

$$(MCl_n)_{sv} + p\,Cl^- \rightleftharpoons [MCl_{n+p}]^{p-}$$

is appreciably on the right side of the equation.

Quantitative measurements on the formation constants of the chloro-complexes have been carried out in benzoyl chloride[27] which has the highest donor number among the solvents under consideration (Table 49). Thus even higher formation constants for chloro-complexes will be expected in thionyl chloride, sulphuryl chloride and nitrosyl chloride.

The formation of chloronium compounds or ionization of acceptor chlorides is hardly expected in a solvent of low donor properties. The only exception appears to be phosphorus(V) chloride, because it gives unsolvated $[PCl_4]^+$ units and thus does not require any stabilization by solvent coordination.

Table 49. *Formation Constants* K *of* $[(C_6H_5)_3C]^+[MCl_{n+1}]^-$
in Benzoyl Chloride at 20°

Acceptor Chloride	Complex formed	K
$SbCl_5$	$[SbCl_6]^-$	$> 10^4$
$GaCl_3$	$[GaCl_4]^-$	$> 10^4$
$FeCl_3$	$[FeCl_4]^-$	$> 10^4$
$SnCl_4$	$[SnCl_5]^-$	300
BCl_3	$[BCl_4]^-$	70
$ZnCl_2$	$[ZnCl_3]^-$	60
$TiCl_4$	$[TiCl_5]^-$	58
$AlCl_3$	$[AlCl_4]^-$	1.5
$SbCl_3$	$[SbCl_4]^-$	1.8
PCl_5	$[PCl_6]^-$	0.2

The chlorides of the alkali- and alkaline earth metals are practically insoluble in the pure liquid solvents, because the low donor numbers do not allow extensive coordination of the cations to take place. Tetraalkylammonium chlorides and phosphorus(V) chloride act as chloride ion donors. Triethylamine, pyridine, quinoline, other N-bases and ketones give conducting solutions which have been interpreted as due to the formation of chloride ions[18]:

$$C_5H_5N + SOCl_2 \rightleftharpoons [C_5H_5NSOCl]^+ + Cl^-$$
$$R_2CO + SOCl_2 \rightleftharpoons [R_2COSOCl]^+ + Cl^-$$

Neutralization reactions have been followed by various techniques. A molybdenum electrode was found useful for potentiometric titrations in anhydrous thionyl chloride[18], but the mechanism of the electrode reaction has not been established. Likewise reactions between sulphur trioxide and various bases have been followed by conductometric and potentiometric titrations in solutions of acetyl chloride[53]. Reactions in thionyl chloride can also be followed by the use of various colour indicators, such as methyl orange, methyl red, phenolphthalein, p-nitrophenol, thymolphthalein or bromine thymolblue, which appear to be indicative of the chloride ion activities of the solutions.

The acidities increase in thionyl chloride[18]: xylenol blue < cresol red < bromothymol blue < bromophenol red < bromophenol blue and in acetyl chloride: cresol red < xylenol blue < bromothymol blue < bromophenol red < bromophenol blue.

It has been concluded from work in acetyl chloride and thionyl chloride that the phenolic hydrogens in the indicator molecules are replaced by solvent cations[54]. This involves the formation of hydrogen chloride when the indicator molecule is dissolved in the oxychloride solvent. Crystal violet may also be used in solutions of thionyl chloride[55].

Iodides are decomposed by thionyl chloride to give free iodine[18]:

$$4 I^- + 2 SOCl_2 \rightarrow S + 2 I_2 + 4 Cl^- + SO_2$$

[53] PAUL, R. C., S. P. NARULA, P. MEYER, and S. K. GONDAL: J. Sci. Ind. Res. **21 B**, 552 (1962); C. A. **58**, 6195 (1963).
[54] GUTMANN, V., and H. HUBACEK: Mh. Chem. **94**, 1019 (1963).
[55] RICE, R. V., S. ZUFFANTI, and W. F. LUDER: Anal. Chem. **24**, 1022 (1952).

Rhenium heptoxide and ammonium perchlorate are solvolysed by thionyl chloride
to give

$$[ReO_3Cl]_2SOCl_3 \text{ and } [NH_4]_2[ReO_2Cl_4 \cdot SOCl_2]$$

respectively[56]. The former gives on heating $ReOCl_4$ which is easily converted into
$[(C_2H_5)_4N]_2[ReCl_6]$ by treatment with tetraethylammonium chloride in thionyl
chloride.

Thionyl chloride is a useful dehydrating agent[57]. Numerous hydrated chlorides
are converted into the anhydrous chlorides by refluxing with thionyl chloride,
since the latter gives with water sulphur dioxide and hydrogen chloride, which are
both volatile.

Likewise benzoyl chloride and benzoyl bromide may be used to dehydrate
hydrated chlorides and bromides[32], when volatile hydrogen halide and benzoic
acid are formed, the latter subliming above $100°$.

Benzoyl bromide is a useful solvent for reactions of bromides. Tetraalkylammonium bromides and most covalent bromides such as $AlBr_3$, $GaBr_3$, $InBr_3$,
$HgBr_2$, $AuBr_3$, $SnBr_4$, $SbBr_3$, $FeCl_3$ and $NbBr_5$ are soluble and give conducting
solutions, whilst bromides of zinc, cadmium, chromium(III), cobalt and nickel
are hardly soluble. Molybdenum electrodes have been found useful to follow
bromide ion transfer reactions potentiometrically. The bromide-ion acceptor
properties were found to increase within a group of the periodic table with
increasing size of the metal, e.g.:

$$AlBr_3 < GaBr_3 < InBr_3$$
$$TiBr_4 < SnBr_4$$
$$NbBr_5 < TaBr_5$$

The same was found by comparing the acidic properties of bromides within one
period

$$HgBr_2 < AuBr_3$$
$$SbBr_3 < SnBr_4 < InBr_3$$
$$WBr_5 < TaBr_5$$

Benzoyl fluoride is a poor solvent for fluorides, but certain reactions leading
to complex fluorides have been shown to occur in its solutions. Tetraalkylammonium fluoride, triethylamine, pyridine and triphenylfluoromethane have basic
properties in anhydrous benzoyl fluoride.

Redox reactions have been observed in nitrosyl chloride, when certain metals
are dissolved,

$$K + NOCl \rightarrow KCl + NO\uparrow$$

and some complex compounds may be obtained by adding an alkali metal to a
solution containing the acceptor chloride, e.g.:

$$[NO]^+[FeCl_4]^- + Na = Na^+[FeCl_4]^- + NO\uparrow$$

Redox reactions in thionyl chloride include the oxidation of titanium(III)
salts[58].

[56] BAGNALL, K. W., D. BROWN, and R. COLTON: J. Chem. Soc. **1964**, 3017.

[57] HECHT, A.: Z. anorg. allg. Chem. **254**, 37 (1947).

[58] FOWLES, G. W. A., and B. J. RUSS: J. Chem. Soc. (A) **1967**, 517.

2. Oxyhalides with Medium Donor Numbers

In the oxyhalides discussed so far the donor properties at the O-atoms are too small to be considered in connection with the formation of anionic complexes and their capacities for solute-dissolution are poor. Solvents of higher donor number allow a wider range of solubilities, since more energy is provided by the interactions with acceptors, such as metal ions. Selenium oxychloride[59], phosphorus oxychloride[8, 60-62] and phenylphosphonic dichloride[8, 60, 62] are much better donor-solvents than the oxyhalides discussed above. Coordination chemistry in their solutions may be explained in terms of competing reactions between the donor solvent molecules and the competitive ligands towards the acceptor molecules or acceptor ions[8, 62].

Table 50. *Some Physical Properties of* $POCl_3$, $SeOCl_2$ *and* $C_6H_5POCl_2$

Property	$POCl_3$	$SeOCl_2$	$C_6H_5POCl_2$
Melting Point (°C)	+1.3	+10	3
Boiling Point (°C)	+108	+178	258
Dielectric Constant at 20°	13.9	46	26
Specific Conductivity at 20° (Ohm^{-1} · cm^{-1})	10^{-8}	10^{-5}	10^{-8}
Donor Number (DN_{SbCl_5})	11.7	12.2	18.5

The solubilities of ionic chlorides are usually small but potassium chloride is soluble in selenium oxychloride and tetraethylammonium chloride is easily soluble in all solvents under consideration. Binary chloro-complexes are usually soluble, but 2:1-electrolytes such as $[R_4N]_2[TiCl_6]$ are nearly insoluble. Most of the covalent chlorides are readily dissolved, including $ZnCl_2$, $CdCl_2$ and $HgCl_2$. Solubilities are considerably enhanced by addition of complex forming agents. For example, although barium chloride is insoluble in pure phosphorus oxychloride, it dissolves in the presence of ferric chloride, because $Ba[FeCl_4]_2$ is formed. Likewise alkali chlorides show higher solubilities in the presence of an acceptor chloride, because alkali chlorometallates are formed.

Considerable heat is evolved when an acceptor chloride is added to the pure oxychloride solvent[63, 64] and numerous solvates have been isolated from the solutions.

Solvates of alkali chlorides, tetraalkylammonium chlorides[65, 66] and pyridine[67] have been described with selenium oxychloride, but claims regarding the preparation of $C_5H_5 \cdot N.POCl_3$ have been disproved[60].

[59] SMITH, G. B. L.: Chem. Revs. **23**, 165 (1938).

[60] GUTMANN, V.: Oesterr. Chem. Ztg. **62**, 326 (1961).

[61] GUTMANN, V.: J. Phys. Chem. **63**, 378 (1959).

[62] BAAZ, M., and V. GUTMANN: Chapter V in "Friedel Crafts and Related Reactions," Ed. G. A. OLAH, Vol. 1, Interscience, New York 1963.

[63] GUTMANN, V., F. MAIRINGER, and H. WINKLER: Mh. Chem. **96**, 574 (1965).

[64] GUTMANN, V., A. STEININGER, and E. WYCHERA: Mh. Chem. **97**, 460 (1966).

[65] WISE, J. R.: J. Amer. Chem. Soc. **45**, 1233 (1923).

[66] AGERMANN, M., L. H. ANDERSSON, I. LINDQVIST, and M. ZACKRISSON: Acta Chem. Scand. **12**, 477 (1958).

[67] LINDQVIST, I., and G. NAHRINGBAUER: Acta Cryst. **12**, 638 (1959).

Table 51. *Stable Solvates and Number of Solvent Molecules of* $POCl_3$,
$C_6H_5POCl_2$ *and* $SeOCl_2$ *Coordinated by One Acceptor Molecule*

Solvent Solute	$POCl_3$	$PhPOCl_2$	$SeOCl_2$
$(C_2H_5)_4NCl$			1, 2, 3, 5
BCl_3	1	1	
$AlCl_3$	1, 2, 6	1, 2	
$GaCl_3$	1, 2	1	
$InCl_3$		1	
$TlCl_3$		1	
$AsCl_3$	1		
$SbCl_3$	1, 2		
$AuCl_3$	1		
$FeCl_3$	1, 1.5		1, 2
$TiCl_4$	1, 2	1, 2	2
$ZrCl_4$	1, 1.5, 2		
$HfCl_4$	1, 1.5, 2		
$SnCl_4$	2	2	2
$TeCl_4$	1		
$SbCl_5$	1, 2	1	1, 2
$NbCl_5$	1		
$TaCl_5$	1		
$MoCl_5$	1		
WCl_6	1 *	1	
$TiOCl_2$	2		
$SnOCl_2$	2		

* unstable at room temperature

GROENEVELD[68] has suggested that coordination occurs through the oxygen
atom of the oxyhalide, and this has been confirmed by the results of X-ray work
on solvates of phosphorus oxychloride[69-75], notably by LINDQVIST and his coworkers, e.g.:

[68] GROENEVELD, W. L.: Rec. Trav. Chim. Pays-Bas **75**, 594 (1956).
[69] LINDQVIST, I., and C. I. BRÄNDÉN: Acta Chem. Scand. **12**, 134 (1958).
[70] LINDQVIST, I., and C. I. BRÄNDÉN: Acta Chem. Scand. **13**, 642 (1959).
[71] BRÄNDÉN, C. I., and I. LINDQVIST: Acta Chem. Scand. **17**, 353 (1963).
[72] BRÄNDÉN, C. I.: Acta Chem. Scand. **16**, 1806 (1962).
[73] BRÄNDÉN, C. I.: Acta Chem. Scand. **17**, 759 (1963).
[74] HERMUDSSON, Y.: Acta Cryst. **13**, 656 (1959).
[75] SHELDON, J. C., and S. Y. TYREE: J. Amer. Chem. Soc. **80**, 4775 (1958).

The same conclusions were drawn from infrared and RAMAN work[75-77]. Data on the chloride ion donor strength of the oxyhalide solvents is, however, not available. One of the main difficulties is the removal of the last traces of water and its elimination during the reactions. Although experimental methods have been considerably improved, it must be born in mind that apparently nobody has ever been successful in working in the complete absence of moisture or of hydrolysis products. Traces of water appear to remain even in reactive liquids, such as the oxychlorides under consideration. It seems that in phosphorus oxychloride small amounts of water form $H_3O^+Cl^-$ which is dissociated in the solution and contrasts with the behaviour of anhydrous hydrogen chloride when dissolved in the same solvent. Thus purified phosphorus oxychloride[78] contains approximately 10^{-4} moles of water per liter and this must be taken into consideration, when the pure solvent or when dilute solutions are considered. The assumption of the self-ionization equilibria in the liquid solvents,[59, 60, 79]

$$SeOCl_2 \rightleftharpoons SeOCl^+ + Cl^-$$
$$POCl_3 \rightleftharpoons POCl_2^+ + Cl^-$$
$$C_6H_5POCl_2 \rightleftharpoons C_6H_5POCl^+ + Cl^-$$

has been regarded as due to chloride ion transfer reactions between solvent molecules,[48, 49]

$$\overset{\displaystyle Cl^-}{\overbrace{SeOCl_2 + SeOCl_2}} \rightleftharpoons SeOCl^+ + SeOCl_3^-$$

and served as a useful basis to account for a number of reactions leading to chlorocomplexes and also for the behaviour of tetraalkylammonium chloride solutions in phosphorus oxychloride during electrolysis. Polymeric $(PO)_n$ is formed at the cathode[80], while chlorine is liberated at the anode in the ratios as required by the reactions[80]:

Cathode: $\qquad 3n[POCl_2]^+ + 3n\varepsilon^- \rightarrow (PO)_n + 2n\,POCl_3$

Anode: $\qquad 3n\,Cl^- \rightarrow \dfrac{3}{2}n\,Cl_2 + 3n\,\varepsilon^-$

An alternative explanation has been offered as most of the reactions found in the solution can be explained without assuming the presence of the self-ionization equilibria[3, 8, 60, 62].

Ionization of dissolved compounds will be discussed with special reference to phosphorus oxychloride, where much data is available. Although ionization of tetraalkylammonium chlorides in this solvent seems to be fairly complete[81], dissociation is incomplete[78, 81] due to the moderate value of the dielectric constant which favours formation of ion pairs and higher associated ionic species even at low concentrations. The FUOSS-function shows a linear characteristic which is an indication of the presence of a binary dissociation equilibrium[81]:

$$(C_2H_5)_4NCl \rightleftharpoons (C_2H_5)_4N^+ + Cl^-; \quad K = 7.10^{-4}$$

[76] WARTENBERG, E. W., and J. GOUBEAU: Z. anorg. allg. Chem. **329**, 269.
[77] GUTMANN, V., and E. WYCHERA: Mh. Chem. **96**, 828 (1965).
[78] GUTMANN, V., and M. BAAZ: Mh. Chem. **90**, 239 (1959).
[79] GUTMANN, V.: Mh. Chem. **83**, 159 (1952).
[80] SPANDAU, H., A. BEYER, and F. PREUGSCHAT: Z. anorg. allg. Chem. **306**, 13 (1960).
[81] GUTMANN, V., and M. BAAZ: Electrochim. Acta **3**, 115 (1960).

The data are interpreted by normal mobilities of the cations and thus do not support the conclusions drawn from transport number measurements[82], for which autocomplex formation in the system $AlCl_3 - POCl_3$ is more likely to be responsible.

Triethylammonium chloride gives in dilute solutions both $[Et_3NH]^+$ and $[Et_3NPOCl_2]^+$ cations in approximately equal amounts[83]. Thus the coordination affinities of the proton and of the supposed $[POCl_2]^+$ ion are similar towards triethylamine[83,84]:

$$[Et_3NH]^+ \ + \ POCl_3 \ \rightleftarrows \ [Et_3NPOCl_2]^+ \ + \ HCl$$
$$-Cl^- \uparrow\downarrow +Cl^- \qquad\qquad\qquad +Cl^- \downarrow\uparrow -Cl^-$$
$$Et_3NHCl \qquad\qquad\qquad\qquad Et_3N+POCl_3$$

Tetraalkylammonium tetrachloroferrate(III) and hexachloroantimonate(V) are electrolytes in phosphorus oxychloride with binary dissociation equilibria. The complex ions are neither solvated nor associated[81,85].

Ionization of a halide ion acceptor such as antimony(V) chloride is explained in acceptor halide solvents (Chapter V), in nitrosyl chloride and in other oxyhalides of low donor numbers as solvents, by halide ion transfer reactions[48,49,85]. This description has also been suggested for phosphorus oxychloride and related solvents[61,86,87] but it is apparent that the solvent-solute interaction is a very decisive factor which has to be taken into consideration. Exchange studies of radio-chlorine between tetraalkylammonium chloride and phosphorus oxychloride as well as selenium oxychloride show rapid exchange[52,88] which is in agreement with the results of the electrochemical investigations[78]. Likewise isotopic exchange reactions occurred instantaneously in solutions in acetyl chloride[89]. With antimony(V) chloride or aluminium chloride very slow exchange was found to occur with the phosphorus oxychloride solvent molecules[52] at $c \sim 10^{-2}M$, but rapid exchange takes place in the presence of chloride ions.

Conductivity studies in solutions of $Cl_5SbOPCl_3$ and of other acceptor solvates in phosphorus oxychloride show the presence of binary dissociation equilibria involving ions with normal mobilities[90]. The equilibrium

$$Cl_5SbOPCl_3 \rightleftharpoons [Cl_4SbOPCl_3]^+ + Cl^-$$

has to be excluded, because this should be shifted to the left side by addition of chloride ions. Conductometric and potentiometric titrations[86,91,92] show that the reaction with tetraethylammonium chloride gives $[(C_2H_5)_4N][SbCl_6]$, which has also been isolated from such solutions. The actual equilibrium is therefore

$$Cl_5SbOPCl_3 \rightleftharpoons [SbCl_6]^- + [POCl_2]^+ \qquad\qquad \text{or}$$
$$2\,Cl_5SbOPCl_3 \rightleftharpoons [SbCl_6]^- + [(POCl_3)_2SbCl_4]^+$$

[82] GUTMANN, V., and R. HIMML: Z. phys. Chem. (N. F.) **4**, 157 (1955).
[83] BAAZ, M., and V. GUTMANN: Mh. Chem. **90**, 276 (1959).
[84] BAAZ, M., and V. GUTMANN: Mh. Chem. **90**, 744 (1959).
[85] GUTMANN, V.: Quart. Revs. **10**, 451 (1956).
[86] GUTMANN, V.: Z. anorg. allg. Chem. **269**, 279 (1952).
[87] GUTMANN, V.: Z. anorg. allg. Chem. **270**, 179 (1952).
[88] LEWIS, J., and D. S. SOWERBY: J. Chem. Soc. **1957**, 336.
[89] FIORANI, M., L. RICCOBONI, and G. SCHIAVON: Bull. Sci. Fac. Chim. Ind. Bologna **21**, 211 (1963); C.A. **60**, 11420 b (1964).
[90] BAAZ, M., and V. GUTMANN: Mh. Chem. **90**, 426 (1959).
[91] GUTMANN, V., and F. MAIRINGER: Z. anorg. allg. Chem. **289**, 279 (1957).
[92] GUTMANN, V., and F. MAIRINGER: Mh. Chem. **89**, 724 (1958).

which may be regarded as ligand exchange reactions, Sb—O → Sb—Cl followed by dissociation. The reaction may better be represented by

$$[SbCl_6]^- + POCl_3 \rightleftharpoons Cl_5SbOPCl_3 + Cl^-; \quad K \sim 10^{-9}$$

Thus chlorine-coordination at the antimony provides a higher stability to the complex species, than oxygen-coordination[90]. The ligand exchange is easily completed when chloride ions are provided to give hexachloroantimonates, while in their absence either the supposed slight self-ionization of the solvent or the presence of unavoidable impurities, such as hydrated hydrogen chloride, are responsible for the occurrence of the reaction. Formation of a chloro-complex in phosphorus oxychloride is therefore different from that in liquid nitrosyl chloride, because the O-donor properties of the solvents are vastly different.

The features of ferric chloride solutions in phosphorus oxychloride merit some detailed discussion. 5.10^{-2} M solutions are dark red[93] and contain polymeric oxygen-coordinated species[94]. Amorphous materials are obtained from such solutions after removal of the solvent[95]. More dilute solutions ($c \sim 10^{-3}$ M) are still red with the main species being a monomeric solvated complex, such as $Cl_3Fe(OPCl_3)_n$ apart from solvated cations and probably tetrachloroferrate ions which contribute to the conductivities of such solutions[95]. From these solutions the compounds $Cl_3FeOPCl_3$ and $(FeCl_3)_2(OPCl_3)_3$ have been obtained. On further dilution with phosphorus oxychloride ($c \sim 10^{-4}$ M) a yellow colour is produced due to the presence of $[FeCl_4]^-$-ions[93,95,96]. The molar conductivities of these solutions are higher than those of the more concentrated solutions. The following equilibria have to be assumed:

$$\left.\begin{array}{l}\text{polymeric anions,} \\ \text{polymeric cations and} \\ \text{undissociated solvate} \\ \text{complex at } c \sim 10^{-3} \text{ M}\end{array}\right\} \rightleftharpoons [FeCl_4]^- + [POCl_2]^+ \quad \text{or} \quad [H_3O]^+ \text{ at } c \sim 10^{-4} \text{ M}$$

The same conclusions are reached from electrolysis experiments in concentrated solutions[97]. MEEK and DRAGO[3] regard the presence of $[FeCl_4]^-$ ions in dilute solutions as due to the occurrence of autocomplex formation as has been found to occur in solutions of ferric chloride in trimethylphosphate[3, 4], e.g.:

$$2\,FeCl_3 + n\,POCl_3 \rightleftharpoons [Cl_2Fe(OPCl_3)_n]^+ + [FeCl_4]^-$$

In this case the spectrum characteristic for the cationic solvated ferric species should be detectable; this is only found either when a stronger chloride ion-acceptor such as antimony(V) chloride is added to the dilute solution of ferric chloride ($c \sim 10^{-4}$ M) due to the reaction:

$$\underset{\text{yellow}}{[FeCl_4]^-} + Cl_5SbOPCl_3 \rightleftharpoons \underset{\text{red}}{Cl_3FeOPCl_3} + [SbCl_6]^-$$

or on concentrating the solution. As may be expected from the medium donor number of phosphorus oxychloride and from the higher stability of the chloro-

[93] GUTMANN, V., and M. BAAZ: Mh. Chem. **90**, 271 (1959).

[94] GUTMANN, V., and F. MAIRINGER: Mh. Chem. **92**, 720 (1961).

[95] GUTMANN, V., M. BAAZ, and L. HÜBNER: Mh. Chem. **91**, 537 (1960).

[96] GUTMANN, V., and M. BAAZ: Mh. Chem. **90**, 729 (1959).

[97] CADE, J. A., M. KASRAI, and I. R. ASHTON: J. Inorg. Nucl. Chem. **27**, 2375 (1965).

complex, replacement of oxygen coordination by chloride coordination can take place in dilute solutions by the chloride ions available in the purified solvent. Ferric chloride gives a yellow solution in *sym.*-tetrachloroethane due to the presence of complex $Cl_3Fe \cdot C_2H_2Cl_4$. This weak coordination is replaced by $POCl_3$ coordination when the latter is added to give the red solutions identical in spectrum with those in phosphorus oxychloride at $c \sim 10^{-3}$ M or in the presence of antimony(V) chloride at $c \sim 10^{-4}$ M. Excess of tetrachloroethane will again replace the O-coordination by $POCl_3$, which is therefore weak.

The situation is different in solutions of aluminium chloride in phosphorus oxychloride, deriving from a higher affinity of aluminium chloride towards O-coordination than in the case of ferric chloride. Polymeric species seem to be present according to the results of precise conductivity measurements[96], the high surface tensions and high viscosities. Conductometric, potentiometric, spectrophotometric and preparative investigations have shown that aluminium chloride reacts in phosphorus oxychloride[98] and in phenylphosphonic dichloride[98] with considerable formation of autocomplex ions. Chloride ions are released by ionization of aluminium chloride due to preferred O-coordination by the solvent. These chloride ions may react with other solvate molecules to give the chloro-complex:

$$\overset{POCl_3}{Cl_3Al(OPCl_3)_n \rightleftharpoons [Cl_2Al(OPCl_3)_{n+1}]^+ + Cl^-}$$
$$[Cl_2Al(OPCl_3)_{n+1}]^+ + POCl_3 \rightleftharpoons [ClAl(OPCl_3)_{n+2}]^{2+} + Cl^-$$
$$\underline{2\,Cl_3Al(OPCl_3)_n + 2\,Cl^- \rightleftharpoons 2[AlCl_4]^- + 2n\,POCl_3}$$
$$3\,Cl_3Al(OPCl_3)_n \rightleftharpoons [ClAl(OPCl_3)_{n+2}]^{2+} + 2[AlCl_4]^- + 2n - 2\,POCl_3$$

The higher affinity for O-coordination of aluminium chloride is responsible for its reaction with ferric chloride, which mainly may be represented for the most part by the equation:

$$Cl_3Al(OPCl_3)_3 + 2\,Cl_3Fe(OPCl_3) \rightleftharpoons [ClAl(OPCl_3)_5][FeCl_4]_2$$

From the solutions a compound of composition $AlCl_3(FeCl_3)_3(POCl_3)_6$ has been isolated, which was suggested to be $[Al(OPCl_3)_6][FeCl_4]_3$ but structural evidence is not available.

Boron(III) chloride gives moderately conducting solutions in phosphorus oxychloride[99] and phenylphosphonic dichloride[100]. Conductivity data and the absence of a reaction with a strong chloride ion donor[99] in phosphorus oxychloride exclude the presence of $[Cl_3POBCl_2]^+Cl^-$ in this solvent, but in $C_6H_5POCl_2$ a reaction with ferric chloride is indicated:

$$Cl_2(C_6H_5)POBCl_3 + FeCl_3 \rightleftharpoons [Cl_2(C_6H_5)POBCl_2][FeCl_4]$$

In a solvent of similar DN_{SbCl_5}, namely diethyl ether, ionization has been formulated as follows[101]:

$$2(C_2H_5)_2OBX_3 \rightleftharpoons [(C_2H_5)_2O)BX_2]^+ + [BX_4]^-$$

[98] BAAZ, M., V. GUTMANN, L. HÜBNER, F. MAIRINGER, and T. S. WEST: Z. anorg. allg. Chem. **311**, 302 (1961).

[99] BAAZ, M., V. GUTMANN, and L. HÜBNER: Mh. Chem. **91**, 694 (1960).

[100] BAAZ, M., V. GUTMANN, and L. HÜBNER: Mh. Chem. **92**, 135 (1961).

[101] SHCHEGOLEVA, T. A., V. D. SHELUDYAKOV, and B. M. MIKHAILOV: Dokl. Akad. Nauk SSSR **152**, 888 (1963); C. A. **60**, 6455 c (1964).

Boron(III) chloride reacts with chloride ion donors in phosphorus oxychloride but no exchange of labelled chlorine is observed between boron(III) chloride and phosphorus oxychloride[102] unless the latter is present in large excess.

Titanium(IV) chloride gives a yellow solution and the spectra are independent from the concentration of the solute. The stepwise replacement of solvent molecules by chloride ions takes place by addition of the latter[94,103]:

$$Cl_4Ti(OPCl_3)_2 + Cl^- \rightleftharpoons [Cl_5TiOPCl_3]^- + POCl_3$$
$$[Cl_5Ti(OPCl_3)]^- + Cl^- \rightleftharpoons [TiCl_6]^{--} + POCl_3$$

When a strong acceptor such as antimony(V) chloride is added $[Cl_3Ti(OPCl_3)_3]^+$ ions are formed according to[103,104]:

$$Cl_4Ti(OPCl_3)_2 + POCl_3 \rightleftharpoons [Cl_3Ti(OPCl_3)_3]^+ + Cl^-$$

Dissociation is expected to occur to a larger extent in selenium oxychloride than in the phosphorus oxyhalides, because the former has a higher dielectric constant and ion pairs or higher ionic aggregations are expected to be present in phosphorus oxychloride solutions. The extent of dissociation will also depend on the nature of the ions: large ions of low charge will be easier dissociated than small ions with high charge.

When moderately concentrated solutions of ferric chloride or aluminium chloride are prepared, it may take several days before the final conductivities are observed. This shows that degradation preceeding dissociation is a slow process.

Chemical association with the formation of chlorine bridges is less likely in a good donor solvent, such as phenylphosphonic dichloride; the solvent molecules may, however, play an important role in stabilizing polymeric structures. Ebullioscopic measurements in solutions of certain acceptor chlorides show higher molecular weights than calculated for monomeric species[94]; the solutions show poor tendencies to crystallization and solvates are precipitated with an amorphous or gel-like character. Various equilibria in concentrated solutions are attained only after prolonged periods. Colloidal solutions are obtained after cooling aluminium chloride solutions in phosphorus oxychloride saturated at a higher temperature. The investigation of the nature of the species and the equilibria involved is a difficult task.

The extent of chloride ion transfer reactions has been extensively studied by spectrophotometric techniques. One of the approaches has been based on the pronounced differences in spectra between solvent-coordinated and fully chloride-coordinated ferric chloride in phosphorus oxychloride[105] and phenylphosphonic dichloride[100]:

$$[R_4N][FeCl_4] + Cl_nMOPCl_3 \rightleftharpoons [R_4N][MCl_{n+1}]^- + Cl_3FeOPCl_3$$

Another method is based on the colourless (unionized) solutions of triphenylchloromethane in the pure solvents[106,107]. Ionization occurs readily in the presence of an acceptor compound[60,106]:

$$(C_6H_5)_3CCl + MCl_n \rightleftharpoons [(C_6H_5)_3C]^+[MCl_{n+1}]^-$$

[102] HERBER, R. H.: J. Amer. Chem. Soc. **82**, 792 (1960).
[103] BAAZ, M., V. GUTMANN, and M. Y. A. TALAAT: Mh. Chem. **92**, 714 (1961).
[104] BRYNTSE, G. A. R., and I. LINDQVIST: Acta Chem. Scand. **14**, 949 (1960).
[105] BAAZ, M., V. GUTMANN, and L. HÜBNER: J. Inorg. Nucl. Chem. **18**, 276 (1961).
[106] BAAZ, M., V. GUTMANN, and J. R. MASAGUER: Mh. Chem. **92**, 590 (1961).
[107] BAAZ, M., V. GUTMANN, and J. R. MASAGUER: Mh. Chem. **92**, 582 (1961).

Tungsten hexachloride, the only hexachloride investigated, accepts one chloride ion from chloride ion donors in phosphorus oxychloride[108]. The complex ion seems to have a coordination number higher than seven due to additional solvent coordination, possibly $[Cl_7WOPCl_3]^-$ but is decomposed by removal of the solvent:

$$[Cl_7WOPCl_3]^- \rightarrow WCl_6 + Cl^- + POCl_3$$

No evidence was found for the formation of $[WCl_8]^{2-}$. Pentachlorides, such as PCl_5, $SbCl_5$, $NbCl_5$, $TaCl_5$ and probably also $MoCl_5$, coordinate one chloride ion to give the hexacoordinated chloro-complexes[60, 62, 92]. Tetrachlorides can coordinate either one or two chlorides ions, e.g. $[Cl_5TiOPCl_3]^-$ and $[TiCl_6]^{2-}$ respectively[60, 62, 92, 103, 109, 110].

All trichlorides accept only one chloride ion to give tetrachlorometallates and dichlorides may coordinate either one or two chloride ions[60, 62].

Table 52. *Coordination Forms of Acceptor-Chlorides in Phosphorus Oxychloride and Phenylphosphonic Dichloride in the Presence of Triphenylchloromethane*

Complex \ Solvent	POCl$_3$	C$_6$H$_5$POCl$_2$
$[PCl_6]^-$	+	
$[PCl_{n<6}]^{p>1-}$	−	−
$[SbCl_6]^-$	+	+
$[SbCl_{n>6}]^{p<1-}$	−	−
$[TiCl_5]^-$	+	+
$[TiCl_{n>5}]^{p<1-}$	−	−
$[SbCl_5]^-$	+	+
$[SnCl_{n>5}]^{p<1-}$	−	−
$[BCl_4]^-$	+	+
$[BCl_{n>4}]^{p<1-}$	−	−
$[AlCl_4]^-$	+	+
$[AlCl_{n>4}]^{p<1-}$	−	−
$[FeCl_4]^-$	+	+
$[FeCl_{n>4}]^{n<1-}$	−	−
$[ZnCl_3]^-$	+	+
$[ZnCl_4]^{2-}$	−	−

Hexachloro- and tetrachlorometallate ions are unsolvated and not associated in solutions. The reluctance of di- and tetrachlorides to accept a second chloride ion from triphenylchloromethane is remarkable, showing that the chloride ion-donor properties of triphenylchloromethane are weak. In solvents of lower donor number such as benzoyl chloride[29] two chloride ions can be accepted from triphenylchloromethane. In the stronger donor solvents, phosphorus oxychloride and phenylphosphonic dichloride, stronger chloride ion donors, such as tetraethylammonium chloride, are neccessary to obtain the fully chloride-coordinated anionic species.

[108] BAAZ, M., V. GUTMANN, and M. Y. A. TALAAT: Mh. Chem. **91**, 548 (1960).
[109] BAAZ, M., V. GUTMANN, M. Y. A. TALAAT, and T. S. WEST: Mh. Chem. **92**, 150 (1961).
[110] BAAZ, M., V. GUTMANN, and T. S. WEST: Mh. Chem. **92**, 164 (1961).

The chloride ion-affinities of acceptor molecules in non-aqueous donor solvents D may be expressed in terms of equilibrium constants for the solvent-chloride ligand exchange reactions in the respective media[62]:

$$Cl_nMD_m + pCl^- \rightleftharpoons [MCl_{n+p}]^{p-} + mD$$

Formation constants are extremely small in solvents of high donor numbers[94] and will increase with decreasing donor properties of the solvent molecules.

Table 53. *Stability Constants of* $[(C_6H_5)_3C][MCl_{n+1}]$ *in Phenylphosphonic Dichloride[107], Phosphorus Oxychloride[106] and Benzoyl Chloride[29] (for Comparison) at Room Temperature*

Chloride	Solvent $C_6H_5POCl_2$ (DN$_{SbCl_5}$ = 18.5)	POCl$_3$ (DN$_{SbCl_5}$ = 11.7)	C_6H_5COCl (DN$_{SbCl_5}$ = 2.3)
ZnCl$_2$	5	12	60
BCl$_3$	11	100	70
AlCl$_3$	0.2	14	1.5
FeCl$_3$	130	290	10,000
TiCl$_4$	5.3	16	58
SnCl$_4$	15.5	85	300
SbCl$_5$	39.4	110	10,000

Table 54. *Approximate Formation Constants* K *of* $[(C_6H_5)_3C][HgCl_3]$-*Complexes in Solvents of Different Donor Numbers*

Solvent	DN$_{SbCl_5}$	K	ε
$(C_4H_9O)_3PO$	25.6	10^{-4}	7
$C_6H_5POCl_2$	18.5	10^{-2}	26
POCl$_3$ ....:.....	11.7	10^{-1}	14
C_6H_5Cl	low	7	5.6

No relationship is found between the dielectric constant and the formation constant; the latter quantity does, however, increase with decreasing donor number of the solvent.

High formation constants of the $[(C_6H_5)_3C][MCl_{n+1}]$ complexes correspond approximately to high $-\Delta H$ values for the formation of the chloro-complexes in a particular solvent using tetraethylammonium chloride as a chloride ion donor.

Table 55. *Heats of Reaction* $(-\Delta H)$ *for the Formation of Chloro-complexes from the Solvates of Acceptor Chlorides and Tetraethyl-ammonium Chloride in Phosphorus Oxychloride[111] and Stability Constants* K *of the Chloro-complexes in Phosphorus Oxychloride at 20°*

Reaction	$-\Delta H$	K
$Cl_3Fe(OPCl_3)_n \rightarrow [FeCl_4]^-$	10.4	290
$Cl_5Sb(OPCl_3)_n \rightarrow [SbCl_6]^-$	9.6	110
$Cl_3B(OPCl_3)_n \rightarrow [BCl_4]^-$	5.7	100
$Cl_4Sn(OPCl_3)_n \rightarrow [SnCl_6]^{--}$	7.2	85
$Cl_4Ti(OPCl_3)_n \rightarrow [TiCl_6]^{--}$	1.1	16

[111] GUTMANN, V., F. MAIRINGER, and H. WINKLER: Mh. Chem. **96**, 828 (1965).

The extent of chloride ion coordination to an acceptor chloride is influenced by the nature of the chloride ion donor, since ionization of triphenylchloromethane does not occur in a donor solvent, but occurs in the presence of an acceptor.

Table 56. *Relative Cl⁻-Acceptor Strength (decreasing from top to bottom) of Acceptor Chlorides towards* $(C_6H_5)_3CCl$ *or* $(p\text{-}CH_3C_6H_4))_3CCl$ *in* $C_6H_5POCl_2$, $POCl_3$ *and* C_6H_5COCl

$C_6H_5POCl_2$	$POCl_3$	C_6H_5COCl
$FeCl_3$	$FeCl_3$	$FeCl_3$
$SbCl_5$	$SbCl_5$	$SbCl_5$
$SnCl_4$	BCl_3	$SnCl_4$
BCl_3	$SbCl_4$	BCl_3
$TiCl_4$	$TiCl_4$	$TiCl_4$
$ZnCl_2$	$AlCl_3$	$ZnCl_2$
$AlCl_3$	$ZnCl_2$	$SbCl_3$
$HgCl_2$	$HgCl_2$	$AlCl_3$
$SbCl_3$	$SbCl_3$	$HgCl_2$

The acceptor strength of strong electron pair acceptors is independent of the properties of the chloride ion donors, but weak electron pair acceptors, such as titanium(IV) chloride or zinc chloride, interact stronger with covalent chloride ion donors than with ionic chlorides. At the same time they accept only one chloride ion from covalent triphenylchloromethane, but two from ionic tetraalkylammonium chloride.

Table 57. *Relative Cl⁻-Acceptor Strength of Chloride Ion-Acceptor Chlorides towards Covalent* $(C_6H_5)_3Cl$ *and Polar* $(C_2H_5)_4NCl$ *Chlorides as Found by Spectrophotometric*[93,95] *and Potentiometric*[109] *Measurements in Phenylphosphonic Dichloride*

Cl- acceptor strength decreasing (from top to bottom) towards	
$(C_6H_5)_3CCl$	$(C_2H_5)_4NCl$
$FeCl_3$	$FeCl_3$
$SbCl_5$	$SbCl_5$
$SnCl_4$	$SnCl_4$
BCl_3	BCl_3
$TiCl_4$	$HgCl_2$
$ZnCl_2$	$TiCl_4$
$AlCl_3$	$AlCl_3$
$HgCl_2$	$ZnCl_2$

The stability of secondary oxonium chlorometallates found,

$$[SbCl_6]^- > [FeCl_4]^- > [AlCl_4]^- > [SnCl_6]^{2-} > [ZnCl_3]^-$$

decreases in the solid state in the same way as that of tertiary oxonium chlorometallates in water[112,113]. The only difference in oxychlorides is the relatively

[112] KLAGES, F., H. MEURESCH, and W. STEPPICH: Ann. Chem. **592**, 81 (1955).

[113] MEERWEIN, H., E. BATTENBERG, H. G. GOLD, E. PFEIL, and G. WILFERT: J. prakt. Chem. **154**, 83 (1940).

stronger acceptor function of aluminium chloride, which has a high coordination affinity towards O- and N-donors also in solutions of benzene or chloroform, for which the following order was found[114]:

$$AlCl_3 > FeCl_3 > SnCl_4 > SbCl_5 > SbCl_3$$

The stabilities of solvates in phosphorus oxychloride and phenylphosphonic dichloride follow the same order:

$$AlCl_3 > FeCl_3$$

Owing to the medium donor strengths of the solvents some weak acceptor chlorides may prefer to allow replacement of one or more chloride positions by the donor solvent molecules. In this way ionization is to occur which should be supported by the presence of very strong chloride ion acceptors, such as antimony(V) chloride or ferric chloride. Such reactions may proceed stepwise in a solvent of donor strength comparable to that of the chloride ion and of reasonable dielectric constant, because replacement of the first position may be easier than that of a second chloride position:

$$MCl_n \rightleftharpoons [MCl_{n-p}]^{p+} + p\,Cl^-$$

Table 58. *Coordination Forms of Acceptor Chlorides in the Presence of* $SbCl_5$ *or* $FeCl_3$ *in Solutions of Phosphorus Oxychloride and Phenylphosphonic Dichloride*

Chloronium Ion \ Solvent	$POCl_3$	$C_6H_5POCl_2$
$[WCl_5]^+$	−	−
$[PCl_4]^+$	+	+
$[PCl_{n<4}]^{p>1+}$	−	−
$[NbCl_4]^+$	+	
$[NbCl_{n<4}]^{p>1+}$	−	
$[TaCl_4]^+$	+	
$[TaCl_{n<4}]^{p>1+}$	−	
$[TiCl_3]^+$	+	+
$[TiCl_{n<3}]^{p>1+}$	−	−
$[SnCl_3]^+$	+	+
$[SnCl_{n<3}]^{p>1+}$	−	−
$[ZrCl_3]^+$	+	
$[ZrCl_{n<3}]^{p>1+}$	−	
$[SeCl_3]^+$	+	
$[SeCl_{n<3}]^{p>1+}$	−	
$[BCl_2]^+$	−	+
$[AlCl_2]^+$	−	−
$[AlCl]^{2+}$	+	+
$[FeCl_2]^+$	−	−
$[FeCl]^{2+}$	−	−
$[ZnCl]^+$	+	+
Zn^{2+}	+	+
$[HgCl]^+$	+	+
Hg^{2+}	−	−

---

[114] HANKE, D. L., and J. STEIGMANN: Anal. Chem. **26**, 1989 (1954).

Such reactions have been followed in phosphorus oxychloride and phenyl-phosphonic dichloride in the presence of chloride ion acceptors such as ferric chloride or antimony(V) chloride. Only the tetrahedral PCl_4^+-ions are non-solvated, but all other chloronium ions seem to be solvated by at least one solvent molecule[8, 60, 62].

Most of the halides produce cations with one unit charge:

$$PCl_5 \rightleftharpoons PCl_4^+ + Cl^-$$

but aluminium chloride gives $[AlCl]^{2+}$ units in the presence of ferric chloride in phosphorus oxychloride[98,105] and phenylphosphonic dichloride[100,109,110]:

$$AlCl_3(POCl_3)_3 + 2\,FeCl_3 \cdot POCl_3 \rightleftharpoons [AlCl(OPCl_3)_5]^{2+} + 2\,[FeCl_4]^-$$

When the solvent is partly removed from a solution containing aluminium chloride and iron(III) chloride in a 1 : 2 molar ratio, a precipitate of $AlCl_3(FeCl_3)_3(POCl_3)_6$ is formed with the possible structure: $[Al(OPCl_3)_6][FeCl_4]_3$.

No chloronium compounds are found in the $FeCl_3$-$POCl_3$ system although their existence has been postulated[3]. Although aluminium chloride and iron(III) chloride have similar chloride ion affinities, the former has a much stronger affinity towards O-coordination than the latter.

Potentiometric and spectrophotometric methods have been used to obtain information about the extent of chloride ion transfer with the formation of chloronium ions and thus about the chloride ion donor strength of metal chlorides in different oxyhalides solvents. The chloride ion donor properties are increased by a solvent of high donor number and decreased by a weak donor solvent, if the chloronium ions are solvated. Thus the chloride ion donor strength of a particular chloride is higher in phenylphosphonic dichloride than in phosphorus oxychloride and is usually non-apparent in benzoyl chloride:

$$C_6H_5POCl_2 > POCl_3 \gg C_6H_5COCl$$

Table 59. *Relative Orders of Chloride Ion-Donor Strength (towards $FeCl_3$) and Chloride Ion-Acceptor Strength (towards $(C_6H_5)_3CCl$) in Phosphorus Oxychloride and Phenylphosphonic Dichloride*[8, 60, 62]

Phenylphosphonic Dichloride		Phosphorus Oxychloride	
Cl⁻ Donor Strength[100]	Reversed Cl⁻ Acceptor Strength[107]	Cl⁻ Donor Strength[115]	Reversed Cl⁻ Acceptor Strength[106]
—	$HgCl_2$	—	$HgCl_2$
$AlCl_3$	$AlCl_3$	$AlCl_3$	$AlCl_3$
$TiCl_4$	$TiCl_4$	$ZnCl_2$	$ZnCl_2$
$ZnCl_2$	$ZnCl_2$	$TiCl_4$	$TiCl_4$
—	—	$HgCl_2$	
BCl_3	BCl_3	BCl_3	BCl_3
$SnCl_4$	$SnCl_4$	$SnCl_4$	$SnCl_4$

The relative order of chloride ion donor strength (or of the relative extent of ionization) is similar in phosphorus oxychloride and in phenylphosphonic dichloride, the only exception being the dichlorides of zinc and mercury. The

[115] BAAZ, M., V. GUTMANN, and L. HÜBNER: Mh. Chem. **92**, 272 (1961).

relative order of chloride ion donor strength is roughly inversely proportional to the chloride ion acceptor strength; this confirms the view, that high interaction with the solvent molecules increases ionization (chloride ion donor properties) and decreases chloride ion acceptor properties.

The order of zinc chloride and titanium(IV) chloride found in phosphorus oxychloride is a reversal of that in phenylphosphonic dichloride, but is reciprocal in the respective solvents. The differences can be accounted for by assuming a solvent contribution, which has opposite effects toward donor and acceptor functions[8, 62]. Zinc chloride has a higher chloride ion donor function in phenylphosphonic dichloride than in phosphorus oxychloride owing to differences in dielectric constant and in solvation by which at the same time its chloride ion acceptor function in phenylphosphonic dichloride is suppressed.

The main reason for the reciprocity of chloride ion donor and chloride ion acceptor functions is associated with the properties of the chlorides. In each pair in Table 60 the stronger chloride ion donor is at the same time the weaker chloride ion-acceptor and *vice versa* irrespective of the solvent. The only exception is zinc chloride, which is stronger than mercuric chloride both as chloride ion donor and as chloride ion acceptor[8, 62].

Table 60. *Relations of Chloride Ion Donor and Chloride Ion-Acceptor Properties of Some Pairs of Chlorides in* $C_6H_5POCl_2$ *and* $POCl_3$

Chloride Ion-Donor Strength	Chloride Ion-Acceptor Strength
$ZnCl_2 > HgCl_2$	$ZnCl_2 > HgCl_2$
$AlCl_3 > BCl_3$	$BCl_3 > AlCl_3$
$AlCl_3 > GaCl_3$	$GaCl_3 > AlCl_3$
$TiCl_4 > SnCl_4$	$SnCl_4 > TiCl_4$
$PCl_5 > SbCl_5$	$SbCl_5 > PCl_5$
$SbCl_3 > SbCl_5$	$SbCl_5 > SbCl_3$

Various colour indicators may be used in oxyhalide systems[17,91,116,117]. Sulphonphthalein indicators were investigated in phosphorus oxychloride[118], thionyl chloride and acetyl chloride[54]. The colour changes occurring reversibly at certain p_{Cl}-values are independent both from the nature of the solvent and from the nature of the solutes changing the chloride ion activity of the solution. The solutions are slightly yellow in media of high chloride ion activity and red in solutions of low chloride ion activities. Red solutions are also formed in highly acidic aqueous solutions.

In agreement with the results of preparative work it was concluded that the solvent cations replace phenolic hydrogens in the indicator molecules on dissolution in phosphorus oxychloride with formation of hydrogen chloride[118]: Colour changes are regarded as due to chloride ion transfer reactions:

$$[Ind(POCl_2)_2]^{2+} + Cl^- \rightleftharpoons [Ind(POCl_2)]^+ + POCl_3$$

---

[116] SINGH, J., R. C. PAUL, and S. S. SANDHU: J. Chem. Soc. **1959**, 845.

[117] PAUL, R. C., J. SINGH, and S. S. SANDHU: J. Ind. Chem. Soc. **36**, 305 (1959); C. A. **54**, 5323 h (1960).

[118] GUTMANN, V., and H. HUBACEK: Mh. Chem. **94**, 1098 (1963).

$$[\mathrm{Ind}(\mathrm{POCl}_2)_2]^{2+} + \mathrm{Cl}^- \rightleftharpoons [\mathrm{Ind}(\mathrm{POCl}_2)]^+ + \mathrm{POCl}_3$$

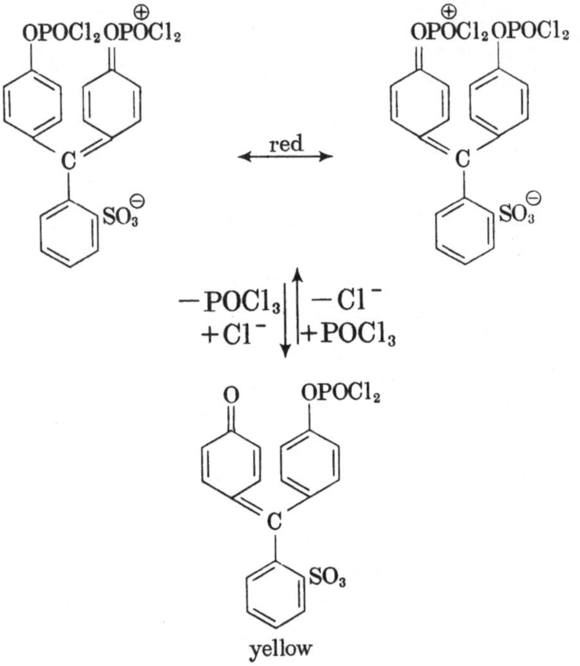

$$\xleftarrow{\quad\text{red}\quad}\rightarrow$$

$$\begin{matrix} -\mathrm{POCl}_3 \\ +\mathrm{Cl}^- \end{matrix} \Bigg\updownarrow \begin{matrix} -\mathrm{Cl}^- \\ +\mathrm{POCl}_3 \end{matrix}$$

yellow

The pink colour of crystal violet is known to be due to resonance between the electron system of the phenyl groups and the free electron pairs of the three dimethylamino-nitrogen atoms. By coordination of one proton the electron pair of one dimethylamino group is removed from the resonance system so that the spectrum of malachite green is produced. By coordination of two further protons the spectrum is changed to that of the triphenyl carbonium ion. The reactions are reversible and known to occur not only in water but also in acetic anhydride[119], acetic acid-dioxane mixtures[120], in thionyl chloride[55], chlorocarbons[117], phosphorus oxychloride[121,122] and acetyl chloride[116].

While in water an enormous excess of protons is neccessary to produce the colour changes, only small amounts of chloride ion acceptors are required in phosphorus oxychloride solutions. The reactions appear to involve the coordination change of the acceptor halides from O-coordination (solvation) to N-coordination, e.g.

$$\mathrm{Cl}_5\mathrm{SbOPCl}_3 + [(\mathrm{CH}_3)_2\mathrm{NC}_6\mathrm{H}_4\mathrm{C}(\mathrm{C}_6\mathrm{H}_4\mathrm{N}(\mathrm{CH}_3)_2)_2]^+ \rightleftharpoons$$
$$\text{pink}$$
$$\rightleftharpoons [\mathrm{Cl}_5\mathrm{Sb}(\mathrm{CH}_3)_2\mathrm{NC}_6\mathrm{H}_4\mathrm{C}(\mathrm{C}_6\mathrm{H}_4\mathrm{N}(\mathrm{CH}_3)_2)_2]^+ + \mathrm{POCl}_3$$
$$\text{green}$$

With antimony(V) chloride the spectrum of malachite green is completed at a molar ratio crystal violet (CV): antimony(V) chloride $= 1:2$ and at a ratio $1:4$ the spectrum of the triphenylcarbonium ion is produced:

[119] SELLÉS, E., and E. S. FLORES: Galenica Acta (Madrid) 8, 291 (1955); C. A. 51, 2234 (1957).
[120] LEVI, L., and O. G. FARMILO: Anal. Chem. 26, 909 (1953).
[121] GUTMANN, V., H. HUBACEK, and A. STEININGER: Mh. Chem. 95, 678 (1964).
[122] GUTMANN, V.: Oest. Chem. Ztg. 65, 273 (1964).

$$[CV]^+Cl^- + 2\,MCl_n \rightleftharpoons [CV(MCl_n)]^+[MCl_{n+1}]^-$$
$$\text{pink} \qquad\qquad\qquad\qquad \text{green}$$
$$[CV(MCl_n)]^+ + 2\,MnCl_n \rightleftharpoons [CV(MCl_n)_3]^+$$
$$\text{green} \qquad\qquad\qquad\qquad \text{yellow}$$

Addition of chloride ions leads to the formation of the more stable chloro-complexes, thus involving the reversed ligand exchange and colour change:

$$[CV(MCl_n)_3]^+ + 2\,Cl^- \rightleftharpoons [CV(MCl_n)]^+ + 2[MeCl_{n+1}]^-$$
$$\text{yellow} \qquad\qquad\qquad\qquad \text{green}$$
$$[CV(MCl_n)]^+ + 2\,Cl^- \rightleftharpoons [CV]^+Cl^- + [MCl_{n+1}]^-$$
$$\text{green} \qquad\qquad\qquad\qquad \text{pink}$$

With weaker acceptor halides colour changes occur at different molar ratios.

Table 61. *Colour Changes of Crystal Violet* (CV) *in Phosphorus Oxychloride in the Presence of Acceptor Halides*

Acceptor Halide	Malachite Green Step at molar ratio $CV : MCl_n$	Triphenyl Carbonium Step at molar ratio $CV : MCl_n$
$SbCl_5$	2.0	4.0
$SnCl_4$	2.0	5.0
$InCl_3$	3.5	7.0
$AlCl_3$	3.0	10.0
BF_3	2.5	13.5
BCl_3	2.6	14.5
$TiCl_4$	8.5	30.0
PCl_5	30.0	?
H^+ in Water	$\sim 10{,}000$	$\sim 10{,}000$

Various bromides and iodides may be solvolysed in $POCl_3$[123], e.g.

$$(C_2H_5)_4NBr + POCl_3 \rightleftharpoons (C_2H_5)_4NCl + POCl_2Br$$

and boron(III) chloride is converted into the fluoride by reaction with phenylphosphonic difluoride[124],

$$2\,BCl_3 + 3\,PhPOF_2 \rightleftharpoons 2\,BF_3 + 3\,PhPOCl_2$$

followed by adduct formation.

Tetraethylammonium permanganate gives, in phosphorus oxychloride, a blue solution which may contain Mn^{3+}. It becomes colourless within a few days at room temperature. The blue solution reacts quantitatively with four moles of ferric chloride and the following interpretation has been suggested[125]:

$$[MnO_4]^- \to [MnCl_8]^- \to MnCl_3 + Cl^- + 2\,Cl_2$$
$$Cl^- + FeCl_3 \to [FeCl_4]^-$$
$$MnCl_3 + 3\,FeCl_3 \to Mn^{3+} + 3\,[FeCl_4]^-$$

Tetraethylammonium chlorate gives a yellow solution in phosphorus oxychloride and reacts in this solution with one equivalent of ferric chloride to give tetraethylammonium tetrachloroferrate[125].

[123] BAAZ, M., and V. GUTMANN: Mh. Chem. **90**, 256 (1959).
[124] GUTMANN, V., J. IMHOF, and F. MAIRINGER: Unpublished.
[125] BAAZ, M., V. GUTMANN, and L. HÜBNER: Mh. Chem. **92**, 707 (1961).

Chapter VII

Certain Donor Solvents

Solvents of low or medium donor numbers are most useful media to carry out ligand exchange reactions and to produce in this way various complex compounds. High donor properties of the solvent although enhancing the solubility properties will lead to high stabilities of the solvate-complexes[1-3].

Water $(DN_{SbCl_5} \sim 18)$ may be considered as a solvent on the border line between solvents of medium and high donor properties and is therefore widely used as a solvent also in coordination chemistry. Due to the high reactivity of the solvent and the availability of the very strongly coordinating hydroxide ions, water will convert many acceptor compounds to hydroxo-complexes and thus prevent the formation of various other complex species.

Many other solvents of medium donor properties are available, such as nitrobenzene $(DN_{SbCl_5} = 4.4)$ and nitromethane $(DN_{SbCl_5} \sim 2.7)$, acetic anhydride $(DN_{SbCl_5} \sim 10.5)$, acetonitrile $(DN_{SbCl_5} = 14.1)$, sulpholane $(DN_{SbCl_5} = 14.8)$, propanediol-1,2-carbonate $(DN_{SbCl_5} = 15.1)$, acetone $(DN_{SbCl_5} = 17.0)$. Alkylacetates and ethers such as diethylether, dioxane or tetrahydrofurane are further examples of donor solvents at the borderline between medium and strong donor properties. With all these solvents the exclusion of water has never completely been achieved and water is playing an important role even if present in trace quantities. Thus donor-acceptor equilibria are influenced by water in for example acetonitrile, which can not be freed from the last traces of water and the purified acetonitrile is still a 10^{-4} M solution of water in this solvent[4].

On the other hand solvents of high donor numbers, such as trialkylphosphates, dimethylformamide, dimethylsulphoxide or pyridine will compete successfully with the donor tendencies of small amounts of water present in their solutions. In fact some of the solvate compounds may even be prepared in aqueous solutions. Wherever it is wished to exclude water as much as possible solvents of high donor numbers should be used, but their use in coordination chemistry is limited to reactions involving strong competitive ligands.

The following examples may illustrate this and some other features of non-aqueous solutions. Apart from dichloroethane and acetonitrile all of the solvents discussed are O-donors.

[1] GUTMANN, V., and E. WYCHERA: Inorg. Nucl. Chem. Letters 2, 257 (1966).
[2] GUTMANN, V., and E. WYCHERA: Rev. Chim. Min. 3, 941 (1966).
[3] GUTMANN, V.: Coord. Chem. Revs., 2, 239 (1967).
[4] GUTMANN, V., and A. STEININGER: Mh. Chem. 96, 1173 (1965).

1. 1,2-Dichloroethane (DN$_{SbCl_5}$ = 0.1)

1,2-Dichloroethane is a poorly coordinating solvent and it has been used as a medium for calorimetric measurements of donor-acceptor reactions[5, 6]. Certain covalent compounds are soluble in 1,2-dichloroethane.

Table 62. *Some Physical Properties of 1,2-Dichloroethane*

Melting Point (°C)	−35.87
Boiling Point (°C)	+83.48
Density at 30° (g · cm^{-3})	1.23831
Viscosity at 30° (Centipoise)	0.730
Dielectric Constant at 20°	10.6
Specific Conductivity at 25° (Ohm^{-1} · cm^{-1}) ...	3 · 10^{-10}
Enthalpy of Vaporization (kcal · mole^{-1})	7.517
Dipole Moment (Debye)	1.75 (benzene)
Donor Number DN$_{SbCl_5}$	0.1
Trouton-Constant (cal · mole^{-1} · deg^{-1})	21.16

Although the donor number is very low, it has been shown, that ferric chloride gives a yellow solution in 1,2-dichloroethane due to a weak interaction between solute and solvent. From this weak complex in solution 1,2-dichloroethane is replaced by addition of a somewhat stronger donor, such as phosphorus oxychloride[7].

1,2-dichloroethane has also been used to follow the complex formation of various acceptor halides, such as $SbCl_5$, $SnCl_4$, $AlCl_3$, $AlBr_3$, $GaBr_3$ and $GaCl_3$ with *trans*-azobenzene, p-chloroazobenzene and p-nitroazobenzene[8, 9]. The structures of such compounds were derived from spectrophotometric and infrared measurements and their formation constants were also given.

It is therefore a useful solvent to study weak complexes as long as the solubilities of the reactants are sufficiently high.

2. Nitromethane (NM) (DN$_{SbCl_5}$=2.7) and Nitrobenzene (NB) (DN$_{SbCl_5}$=4.4)

Table 63. *Some Physical Properties of NM and NB*

	NM	NB
Melting Point (°C)	−28.5	+5.8
Boiling Point (°C)	+101	+210.8
Density at 25° (g · cm^{-3})	1.130	1.193
Viscosity at 30° (Centipoise)	0.612	1.634
Dielectric Constant at 30°	35.9	34.8
Specific Conductivity at 25° (Ohm^{-1} · cm^{-1})	10^{-7}	10^{-7}
Donor Number DN$_{SbCl_5}$	2.7	4.4
Enthalpy of Vaporization (kcal · mole^{-1})	8.126	9.749
Dipole Moment (Debye)	3.54 (gas)	4.23 (gas)
Trouton-Constant (cal · mole^{-1} · deg^{-1})	21.71	20.14

[5] GUTMANN, V., A. STEININGER, and E. WYCHERA: Mh. Chem. **97**, 460 (1966).
[6] OLOFSSON, G., I. LINDQVIST, and S. SUNNER: Acta Chem. Scand. **17**, 259 (1963).
[7] BAAZ, M., and V. GUTMANN: Mh. Chem. **91**, 537 (1960).
[8] STEININGER, A., and V. GUTMANN: Mh. Chem. **97**, 171 (1966).
[9] GUTMANN, V., and U. MAYER: Mh. Chem. **98**, 294 (1967).

Nitromethane and nitrobenzene are useful solvents with weak donor properties. Nitromethane is difficult to purify and nitrobenzene absorbs in the visible range. Tetraalkylammonium halides are somewhat soluble and the solutions conduct electricity, since reasonable dissociation is encouraged by the high dielectric constants of the solvents[10].

Acceptor halides are usually soluble and solvates have been obtained of $SbCl_5$, $AlCl_3$, $AlBr_3$, $TiCl_4$, $TiBr_4$, TiF_4, $CaCl_2$, $MnCl_2$, $CdCl_2$ and $ZnCl_2$. Conductivity studies indicate interaction with $SnCl_4$, $AsCl_3$ and $SbCl_3$. The conductivities of the solutions decrease $SbCl_5 > SnCl_4 > TiCl_4 > AsCl_3$ and for protonic acids $HSO_3F >$ $> H_2SO_4 > HCOOH > CH_3COOH$[11]. The existence of self-ionization equilibria in the pure liquids has also been postulated[11]:

$$CH_3NO_2 \rightleftharpoons CH_2NO_2^+ + H^+$$
$$C_6H_5NO_2 \rightleftharpoons C_6H_4NO_2^+ + H^+$$

Solubilities of transition metal halides are frequently low. The solubilities of nickel chloride and cobalt chloride[12] are $NiCl_2 < CoCl_2$ and are known to increase in general with increasing donor number of the solvent:

$$C_2H_4Cl_2 < NB < AN < TMP < DMF < DMSO$$

Cobalt(II)chloride gives a tetrahedral disolvate with nitromethane[12] with properties which are completely different from those of the dimethylformamide compounds[13] in that the solutions in nitromethane[13] are non-conducting.

$$CoCl_2 + 2\,NM \rightleftharpoons CoCl_2(NM)_2$$

Chloro-complexes namely $[CoCl_3]^-$ and $[CoCl_4]^{2-}$ are easily formed in the presence of chloride ion donors[13a, 13b]. Likewise $CoBr_2$, CoI_2, $Co(N_3)_2$ and $Co(NCS)_2$ are non-electrolytes in NM and $Co(CN)_2$ is scarcely soluble in this solvent. $[CoBr_3]^-$, $[CoBr_4]^{2-}$, $[CoI_3]^-$, $[CoI_4]^{2-}$, $[Co(N_3)_4]^{2-}$, $[Co(NCS)_4]^{2-}$, $[Co(CN)_4]^{2-}$ and $[Co(CN)_5]^{3-}$ are formed quantitatively in the presence of the stoichiometrically required amounts of the respective halide or pseudohalide ion[13a, 13b]. Analogous behaviour is found for the VO^{2+} ion [13c].

Absorption spectra and conductance data have been reported for anhydrous ferric chloride solutions in pure nitromethane[14]. The data have been explained by supposing that one of the two process is to occur[14]:

$$Fe_2Cl_6 \rightleftharpoons \text{complex} \rightleftharpoons [FeCl_4]^- + [FeCl_2]^+ \quad \text{or}$$
$$Fe_2Cl_6 \rightleftharpoons \text{complex} \rightleftharpoons [Fe_2Cl_5]^+ + Cl^-$$

From the reddish brown fluorescent solutions of ferric chloride at $10^{-2}\,M$ a flocculent reddish brown precipitate of unknown composition is formed and BEER's law is not obeyed not even in dilute solutions[14]. It seems therefore that polymeric

[10] DRAGO, R. S., and K. F. PURCELL: Chapter 5 "Non-Aqueous Solvent Systems," Ed. T. C. WADDINGTON, Academic Press, London, New York 1965.

[11] PAUL, R. C., R. KAUSHAL, and S. S. PAHIL: J. Ind. Chem. Soc. **42**, 483 (1965).

[12] COTTON, F. A., and R. S. HOLM: J. Amer. Chem. Soc. **82**, 2979 (1960).

[13] BUFFAGNY, S., and T. M. J. DUNN: J. Chem. Soc. **1961**, 5105.

[13a] GILL, N. S., and R. S. NYHOLM, J. Chem. Soc. **1959**, 3997.

[13b] GUTMANN, V., and K. H. WEGLEITNER,: Mh. Chem. **99**, 368 (1968).

[13c] GUTMANN, V., and H. LAUSSEGGER: Mh. Chem. in press (1968).

[14] DE MAINE, P. A. D., and E. KOUBEK: J. Inorg. Nucl. Chem. **11**, 329 (1959).

species may be present in the solutions, as have been found in solutions of ferric chloride in phosphorus oxychloride[15], which has somewhat better donor properties than nitromethane[2, 3].

Numerous systems involving mainly halides of the representative elements have been investigated in nitrobenzene mainly by electrochemical methods. Examples of more recent work are the investigations of the systems $SnBr_4$—$AlBr_3$ and $SbCl_5$—$AlCl_3$[16] where no compound formation was found and the system $AlBr_3$—THF[17] where a $1:2$ adduct was shown to be formed.

Nb_2Cl_{10} is dimeric in solutions of nitromethane[18] or carbon tetrachloride, since breakdown of the chlorine bridges requires more energy, than can be provided by the interaction with the solvents but it is monomeric in stronger donor solvents such as acetonitrile[18].

Triphenylchloromethane gives in nitromethane the yellow spectrum of the triphenyl carbonium ion[19], which is known to be formed in the presence of acceptor compounds. POCKER[20] has shown that actually hydrogen chloride is present in such solutions, which may therefore be responsible for the ionization of triphenylchloromethane in nitromethane.

It may be concluded that nitromethane is an extremely useful solvent for the formation of various complex compounds.

3. Acetic Anhydride (AA) (DN$_{SbCl_5}$ = 10.5)

The solvent properties of acetic anhydride have been reviewed in considerable detail[21]. Alkali acetates, acetyl halides, alkali iodides, halides of phosphorus(III), arsenic(III), antimony(V) and other acceptor compounds as well as many other covalent compounds are usually soluble in acetic anhydride.

Table 64. *Some Physical Properties of Acetic Anhydride*

Melting Point (°C)	−73.0
Boiling Point (°C)	+140.0
Density at 20° (g · cm^{-3})	1.08107
Viscosity at 25° (Centipoise)	0.85
Dielectric Constant at 20°	22.1
Specific Conductivity at 25° (Ohm^{-1} · cm^{-1})	$5 \cdot 10^{-7}$
Dipole Moment (Debye)	2.8 (gas)
Donor Number DN$_{SbCl_5}$	10.5
Enthalpy of Vaporization (kcal · mole^{-1})	9.393
Trouton-Constant (cal · mole^{-1} · deg^{-1})	22.86

[15] GUTMANN, V., and M. BAAZ: Mh. Chem. **90**, 271 (1959).

[16] GORENBEIN, E. YA., V. V. SUKHAN, and I. L. ABARBARCHUK: Ukrain. Khim. Zh. **29**, 797 (1963); C. A. **60**, 1324 d (1964).

[17] GORENBEIN, E. YA., and G. G. RUSIN: Zh. Neorg. Khim. **10**, 458 (1965); C. A. **62**, 3453 b (1965).

[18] KEPERT, D. L., and R. S. NYHOLM: J. Chem. Soc. **1965**, 2871.

[19] BENTLEY, A., A. G. EVANS, and J. HALPERN: Trans. Farad. Soc. **47**, 711 (1951).

[20] POCKER, Y.: J. chem. Soc. **1958**, 240.

[21] SURAWSKI, H.: Part 2 Vol. IV, "Chemistry in Non Aqueous Ionizing Solvents," Ed. G. JANDER, H. SPANDAU, C. C. ADDISON, p. 132, Vieweg and Interscience, Braunschweig, London and New York 1963.

Ionic reactions occurring in acetic anhydride have been explained by assuming the existence of a self-ionization equilibrium[22]:

$$(CH_3CO)_2O \rightleftharpoons CH_3CO^+ + CH_3COO^-$$

The conductivities of acetyl halides are very low, but they can react with alkali acetates to give the insoluble alkali halides

$$CH_3COCl + CH_3COOK = KCl \downarrow + (CH_3CO)_2O$$

The reactions were followed by conductometric and potentiometric titrations[22]. Amperometric titrations were also carried out[23].

The donor properties of acetic anhydride are similar to those of phosphorus oxychloride discussed in the previous chapter and many adducts are known to be formed with acceptor compounds, such as the high melting $SnCl_4(AA)_2$ [24] as well as $TiCl_4(AA)_2$, $SO_3 \cdot AA$, $SbCl_5 \cdot AA$ [25, 26], $MoCl_5 \cdot AA$, $FeCl_3 \cdot AA$, $AlCl_3 \cdot AA$ and $TeCl_4(AA)_2$ [27]. No significant shifts of the $C=O$ i.r. band occur on complex formation with stannic chloride[24].

It is to be expected that such compounds can easily be converted into the corresponding chloro-complexes by addition of chloride ion-donors, but such experiments have not been carried out yet in acetic anhydride.

Lewis bases, such as pyridine, quinoline and α-picoline give also 1 : 1 adducts. These compounds have been formulated as ionic compounds, e.g.

$$[C_5H_5NCH_3CO]^+[CH_3COO]^-$$

but structural evidence has not been given. Reactions between acceptor compounds and nitrogenous bases have been followed in anhydrous acetic anhydride by colour indicators such as crystal violet, malachite green or benzanthrene[27]. They were interpreted as supporting the self-ionization of acetic anhydride[27] but the reaction products have not been isolated and identified. The reactions may just as well be considered as replacement reactions, by which the weaker donor AA is replaced by the stronger donor N-base in the coordination compound:

$$X_nMO(COCH_3)_2 + C_5H_5N = X_nMC_5H_5N + (CH_3CO)_2O$$

Protonic acids, such as HSO_3F, HSO_3Cl, Cl_2HCOOH give solid adducts, which can be titrated with bases in acetic anhydride, using colour indicators[28]. Oxides, carbonates, oxalates, formates, sulphites and nitrates of many elements are converted into the corresponding acetates by acetic anhydride[29].

With phosphorus(V) chloride a vigorous solvolysis reaction takes place according to the equation[30]:

$$(CH_3CO)_2O + PCl_5 \rightarrow 2\,CH_3COCl + POCl_3$$

[22] JANDER, G., E. RÜSBERG, and H. SCHMIDT: Z. anorg. allg. Chem. **255**, 238 (1948).

[23] GUTMANN, V., and E. NEDBALEK: Mh. Chem. **89**, 203 (1958).

[24] HUNT, P., and D. P. N. SATCHELL: J. Chem. Soc. **1964**, 5437.

[25] USANOVICH, M., and K. YATSIMIRSKI: Zh. Obshch. Khim. **11**, 954 (1941); C. A. **36**, 6444 (1942).

[26] USANOVICH, M., and K. YATSIMIRSKI: Zh. Obshch. Khim. **11**, 959 (1941); C. A. **36**, 6444 (1942).

[27] PAUL, R. C., K. C. MALHOTRA, and O. C. VAIDYA: Ind. J. Chem. **3**, 1 (1965).

[28] PAUL, R. C., K. C. MALHOTRA, and K. C. KHANNA: Ind. J. Chem. **3**, 63 (1965).

[29] PAUL, R. C., K. C. MALHOTRA, and O. C. VAIDYA: Ind. J. Chem. **3**, 97 (1965).

[30] WALE, C. L.: Chem. Ind. **1963**, 2003.

The reactions

$$MCl_n + n\,LiClO_4 \rightleftharpoons M(ClO_4)_n + n\,LiCl$$

were studied by the potentiometric method using Ag/Ag-acetate electrodes in 0.1 M solutions of $LiClO_4$ in acetic anhydride[31].

Iodine was found to be reduced electrochemically in acetic anhydride at a platinum electrode to give the triiodide ion[32]. Electrolytic studies in acetic anhydride have shown that antimony(III) bromide and arsenic(III) chloride are reduced to the elements, while mercuric bromide is reduced to Hg_2Br_2 at the cathode[33].

4. Acetonitrile (AN) (DNSbCl$_5$ = 14.1)

Acetonitrile is a very versatile solvent[34]. It covers a convenient liquid range, is easy to handle and has a relatively high dielectric constant. The solvent molecule is highly polar and rod-like in shape. The medium donor number[1] of 14.1 makes it an excellent solvent for numerous classes of compounds and allows at the same time the occurrence of a large number of ligand-exchange reactions[3]. It has been concluded from magnetic succeptibility measurements that acetonitrile may be classified as a soft potentially π-bonding ligand.

Although its donor number is smaller than that of phenylphosphonic dichloride[1] discussed in the previous chapter, it has a high tendency to solvate metal ions and acceptor molecules. Undoubtedly the favourable steric properties contribute considerably to the high tendency to coordinate metal ions. Thus numerous classes of compounds are soluble and ionized and the high dielectric constant allows considerable dissociation to take place. Accordingly a large number of ionic reactions can be carried out in its solutions.

Table 65. *Physical Properties of Acetonitrile*

Melting Point (°C)	−45.7
Boiling Point (°C)	81.6
Density (g · cm^{-3})	0.7767
Viscosity (Centipoise) at 30°	0.3448
Donor Number DN$_{SbCl_5}$	14.1
Dielectric Constant at 20°	38.8
Specific Conductivity at 20° (Ohm^{-1} · cm^{-1})	10^{-8}
Enthalpy of Vaporization (kcal · mole^{-1})	7.827
Dipole Moment (Debye)	3.2
Trouton-Constant (cal · mole^{-1} · deg^{-1})	22.09

Acetonitrile shows a higher proton affinity than water which behaves as a base in its solutions[35]. Perchloric acid behaves as a strong acid in acetonitrile[36],

[31] BADOZ-LAMBLING, J., and V. PLICHON: Bull. Soc. Chim. France **1963**, 1014.

[32] PLICHON, V., J. BADOZ-LAMBLING, and G. CHARLOT: Bull. Soc. Chim. France **1964**, 287.

[33] SCHMIDT, H., I. WITTKOPF, and G. JANDER: Z. anorg. allg. Chem. **256**, 113 (1948).

[34] WALTON, R. A.: Quart. Revs. **19**, 126 (1965).

[35] KOLTHOFF, I. M., S. BRUCKENSTEIN, and M. K. CHANTOONI, Jr.: J. Amer. Chem. Soc. **83**, 3927 (1961).

[36] COETZEE, J. F., and I. M. KOLTHOFF: J. Amer. Chem. Soc. **79**, 6110 (1957).

which does not show any marked tendency for hydrogen bonding. Thus aceto-nitrile is a solvent of very low proton-donating (acidic) tendencies, with a difference in basicity compared with water of 5 orders of magnitude.

On the other hand it has similar, although somewhat weaker, donor properties than water. Like the latter acetonitrile has favourable steric characteristics for the formation of solvated metal ions and of other donor-acceptor compounds; the nitrogen atom of the solvent molecule acts as the donor atom[34].

Complex formation occurs with Lewis acids[34], such as $SbCl_5$, $SnCl_4$, $TiCl_4$, BCl_3, $AlCl_3$, $GaCl_3$, $InCl_3$, BBr_3, $AlBr_3$, $GaBr_3$ and BF_3. Ligand exchange reactions with trans-azobenzene, p-dimethylaminoazobenzene and p-methoxyazobenzene have been studied by the spectrophotometric method in acetonitrile[4]. The spectra were found to be independent from the nature of the acceptor and nearly identical with those of the protonated azo-compounds. The strong acceptor antimony(V) chloride, when present in a large excess in acetonitrile solutions containing the potential multidentate donor p-dimethylaminoazobenzene, is found to coordinate at one azo nitrogen atom and the nitrogen atom of the amine group. The equilibria are considerably influenced by the presence of even small amounts of water, which is a stronger donor than acetonitrile[4]:

$$BF_3 \cdot CH_3CN + H_2O \rightleftharpoons BF_3 \cdot OH_2 + CH_3CN; \qquad K \sim 10^{-3}$$
$$SbCl_5 \cdot CH_3CN + H_2O \rightleftharpoons SbCl_5 \cdot OH_2 + CH_3CN; \qquad K \sim 10^{-4}$$

It is impossible to mention all of the solvate-complexes of acetonitrile and reference to some of these compounds will be made later. Most of the transition metal ions give hexasolvated species.

The solvation energy is high enough to dissolve numerous compounds, but many ligands can easily replace coordinated solvent molecules. Ligand exchange reactions of the type

$$BCl_3 \cdot AN + L \rightleftharpoons BCl_3L + AN$$

have been carried out[37]. This gives the donor order: pyridine > tetrahydrofurane > > AN.

Triphenylchloromethane, which is not ionized in the pure solvent, has been used as a source of chloride ions to study the formation of chloro-complexes of various acceptor chlorides[38]. It has been shown that ionization of triphenylchloro-methane is only promoted by acceptor compounds[39]. Triphenylcarbonium hexa-chloroantimonate was shown to have an ionic structure in the crystal with a "propeller-structure"[40] for the cation.

It has been found that most of the acceptor chlorides accept one chloride ion from triphenylchloromethane in a solution of acetonitrile, including the tetra-chlorides of titanium(IV) and tin(IV). Zirconium(IV) and germanium(IV)-chlorides were, however, found to accept two chloride ions to produce the hexa-chlorometallates. A careful analysis of the data has revealed that inner-sphere complexes are formed in which the undissociated ion pairs are not separated by solvation shells[38]:

$$(C_6H_5)_3CCl + MCl_n \rightleftharpoons [(C_6H_5)_3C]^+[MCl_{n+1}]^-$$

[37] GERRARD, W., M. F. LAPPERT, and J. W. WALLIS: J. Chem. Soc. 1960, 2178.
[38] BAAZ, M., V. GUTMANN, and O. KUNZE: Mh. Chem. 93, 1142 (1962).
[39] LICHTIN, N. N., and P. D. BARTLETT: J. Amer. Chem. Soc. 73, 5530 (1951).
[40] SHARP, D. W. A., and N. SHEPPARD: J. Chem. Soc. 1957, 674.

No measurable electrostatic dissociation is found to take place and the molar conductivities found in the solutions may be due to the presence of ionized acceptor solvates which have not been converted into the chloro-complexes. The conductivity of a solution of antimony(V) chloride is explained according to the dissociation reaction[41, 42]:

$$2\,Cl_5Sb \cdot NCCH_3 \rightleftharpoons [Cl_4Sb(NCCH_3)_2]^+ + [SbCl_6]^-$$

Titanium(IV) chloride accepts also one chloride ion, but analysis of the data shows the formation of a dimeric anion[38]:

$$2\,TiCl_4(AN)_2 + 2\,Cl^- \rightleftharpoons [Ti_2Cl_{10}]^{--} + 4\,AN$$

which has probably a chlorine bridged structure:

$$\left[Cl_4Ti \diagup^{\displaystyle Cl}_{\diagdown Cl}\diagdown^{\diagup} TiCl_4 \right]^{2-}$$

The results (Table 66) show that, owing to the medium donor number of the solvent, the acceptor strengths are differentiated in acetonitrile because of the reciprocal effects between the strength of solvent coordination and formation constant of the halide complex[43].

Table 66. $\log K$ (Formation Constants) for Chloro-complexes $[(C_6H_5)_3C]^+[MCl_{n+1}]^-$ Formed from Different Acceptor Chlorides in Acetonitrile $(DN_{SbCl_5} = 14.1)$ and Phenylphosphonic Dichloride $(DN_{SbCl_5} = 18.5)$ [1, 43]

Acceptor Chloride	AN	$C_6H_5POCl_2$
$ZnCl_2$	1.84	0.72
$HgCl_2$	1.64	−0.7
BCl_3	1.32	1.03
$AlCl_3$	1.75	−0.72
$GaCl_3$	4.8	—
$SbCl_3$	0.7	−1.6
$SnCl_4$	4.3	1.2
$TiCl_4$	1.9	0.7
$SbCl_5$	5.1	1.6

Thus all chloro-complexes show higher formation constants in acetonitrile than in phenylphosphonic dichloride. It may be noted that boron(III) chloride is a stronger chloride ion acceptor than aluminium chloride in oxyhalide systems; this order is reversed in acetonitrile, where aluminium chloride accepts chloride ions more readily than in oxyhalides. This shows once more the preferred coordination of aluminium chloride with oxygen donors, which also reflects its strong interaction with ketones[44].

[41] KOLDITZ, L., and H. PREISS: Z. anorg. allg. Chem. **310**, 242 (1961).
[42] BEATTIE, I. R., and M. WEBSTER: J. Chem. Soc. **1963**, 38.
[43] BAAZ, M., and V. GUTMANN: Chapter V in "Friedel Crafts and Related Reactions," Ed. G. A. OLAH, Interscience, New York, London 1963.
[44] SUSZ, B. P., and P. CHALONDON: Helv. Chim. Acta **41**, 1332 (1958).

The formation constants of chloro-complexes in acetonitrile are of the same order of magnitude as they are in water. While the order of chloro-complex formation in water is

$$HgCl_2 > ZnCl_2$$

it is reversed in acetonitrile

$$ZnCl_2 > HgCl_2$$

showing the well known high affinity of mercury(II), which is a class (b) metal ion[45], towards N-donors.

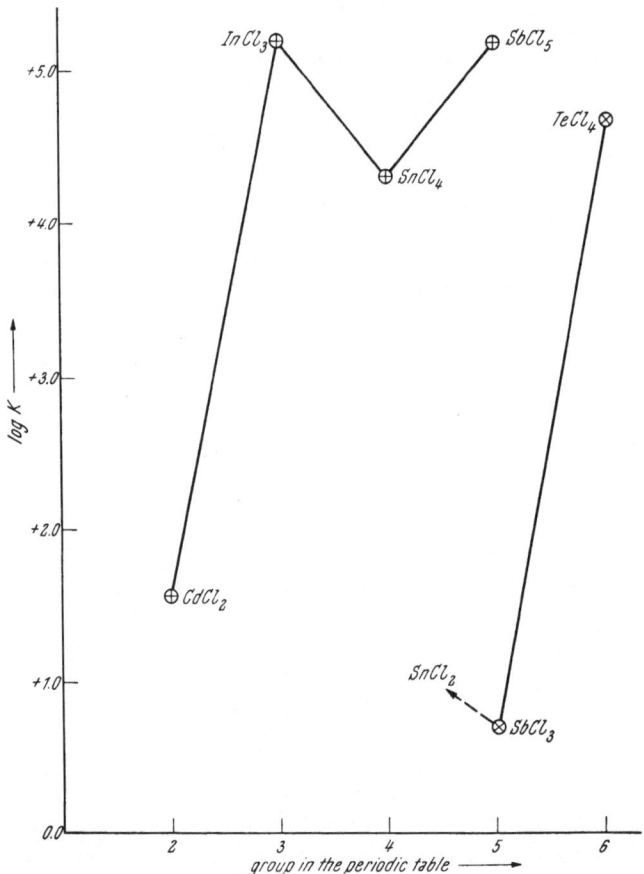

Fig. 10. Formation constants of chloro-complexes from $(C_6H_5)_3CCl$ and certain chlorides in acetonitrile

Interpretation of the results shown in Fig. 10 suggested the following contributions to the understanding of chloride ion-acceptor properties of chlorides of the representative elements[43, 46].

[45] IRVING, H. M. N. H.: Chem. Soc., Spec. Publ. No. 13, 13 (1959).
[46] BAAZ, M., V. GUTMANN, and O. KUNZE: Mh. Chem. 93, 1162 (1962).

a) Ligand effect: Chlorides and chloro-complexes without unshared electron pairs and with an odd number of ligands (three or five) are strong chloride ion-acceptors ($GaCl_3$, $FeCl_3$, $SbCl_5$, $[SnCl_5]^-$) while those with an even number (two or four) of ligands (again without unshared electron pairs) are weak chloride ion-acceptors ($GeCl_4$).

b) Ligand effect of the unshared electron pair: Unshared electron pairs, which may be considered as pseudo ligands in structural considerations appear to behave as pseudo ligands in complex forming reactions in acetonitrile. Tetrachlorides with one lone electron pair, such as tellurium(IV) chloride, as well as trichlorides with two unshared electron pairs, such as iodine(III) chloride, will act as pseudo-pentachlorides, which according to the ligand effect have strong chloride ion affinities. A trichloride which behaves as a pseudotetrachloride due to the presence of one lone electron pair such as $AsCl_3$ or $SbCl_3$ will be much weaker as a chloride ion acceptor than gallium(III) chloride or indium(III) chloride because it is to be considered as a pseudotetrachloride. To account for the ligand effect (a) the sum of the number of ligands and pseudoligands (unshared electron pairs) must be taken into consideration[47].

c) Screening effect of the pseudo-ligand (unshared electron pair): A pseudo-n-chloride is a weaker chloride ion acceptor than the corresponding true n-chloride, since the contribution to the chloride ion affinity of the molecule or ion provided by a pseudo-ligand is smaller than by a real ligand ($TeCl_4 < SbCl_5$ and $SbCl_3 < SnCl_4$).

d) Charge effect: In some solvents of low or medium dielectric constant the formation of free ions is depressed. Some di- and tetrahalides may not form tetra- and hexachlorometallates respectively, but prefer to reach the stable coordination number by dimerization with the formation of chlorine bridges. Thus $[TiCl_5]^-$ is dimeric in acetonitrile solution:

$$2\,TiCl_4 + 2\,Cl^- \xrightarrow{\text{Acetonitrile}} [Ti_2Cl_{10}]^{2-}$$

In the presence of unshared electron pairs screening and charge effects may suppress the ligand effect of the unshared electron pair.

e) Size effect: Increase in size of the central atom of the acceptor chloride increases the acceptor strength, but suppresses the ligand effect (a) and the charge effect (d). $BCl_3 < AlCl_3 < GaCl_3 < InCl_3 > TlCl_3$. As can be seen from this example the affinity maximum is found for chlorides in the fifth period of the periodic table.

The superposition of these effects may account for the observed formation constants of single charged triphenylchloromethane-complexes in acetonitrile:

$K > 10^5$: $InCl_3$, $SbCl_5$, $GaCl_3$, ICl_3

K: 10^5 to 10^4: $TeCl_4$, $SnCl_4$

K: 10^4 to 10^3: $PbCl_4$, $TaCl_5$, $NbCl_5$

K: 10^3 to 10^2: $TlCl_3$

K: 10^2 to 10^1: $TiCl_4$, $ZnCl_2$, $AlCl_3$, $HgCl_2$, $CdCl_2$, BCl_3

K: 10^1 to 1: $BiCl_3$, $SbCl_3$, $GeCl_4$, PCl_5

$K < 1$: $AsCl_3$, $SiCl_4$, PCl_3, $\langle CCl_4 \rangle$

[47] MUETTERTIES, E. L., and R. A. SCHUNN: Quart. Revs. **20**, 245 (1966).

It has been shown that in phosphorus oxychloride, which is somewhat smaller in donor number than acetonitrile, certain chlorides are able to act as chloride ion donors to give chloronium ions. Thus phosphorus(V) chloride gives PCl_4^+ ions:

$$PCl_5 \rightleftharpoons PCl_4^+ + Cl^-$$

Polyhalides are ionized in acetonitrile[48]:

$$IPCl_6 \rightleftharpoons PCl_4^+ + ICl_2^-$$
$$AlX_3(CH_3CN)_n \rightleftharpoons [AlX_m(CH_3CN)_n]^{+3-m} + (3-m)X^-$$

and the tetrachloroiodate(III) ion ICl_4^-, which is well established in the solid state, is stabilized by the trimethylsulphonium ion in acetonitrile solution[49].

In acetonitrile vanadium(III) chloride[50] gives $VCl_3 \cdot (AN)_3$, titanium(III) chloride[50] gives $TiCl_3(AN)_3$ and chromium(III) chloride $CrCl_3(AN)_3$. These complexes in acetonitrile solution were all found to act as chloride ion donors in the presence of strong chloride ion acceptors[51]:

$$VCl_3 + 3\,SbCl_5 \rightleftharpoons V_{sv}^{3+} + 3\,[SbCl_6]^-; \qquad K_1 \sim 10^2$$
$$2\,VCl_3 + 3\,SnCl_4 \rightleftharpoons 2\,V_{sv}^{3+} + 3\,[SnCl_6]^{2-}; \qquad K_2 \sim 5 \cdot 10^{-1}$$
$$2\,VCl_3 + 3\,TiCl_4 \rightleftharpoons 2\,V_{sv}^{3+} + 3\,[TiCl_6]^{2-}; \qquad K_3 \sim 5 \cdot 10^{-7}$$
$$CrCl_3 + SbCl_5 \rightleftharpoons [CrCl_2]^+ + [SbCl_6]^-; \qquad K_4 \sim 5 \cdot 10^2$$
$$2\,CrCl_3 + SnCl_4 \rightleftharpoons 2\,[CrCl_2]^+ + [SnCl_6]^{2-}; \qquad K_5 \sim 5 \cdot 10^1$$
$$2\,CrCl_3 + TiCl_4 \rightleftharpoons 2\,[CrCl_2]^+ + [TiCl_6]^{2-}; \qquad K_6 \sim 10^{-1}$$

While no chloronium compounds of vanadium(III) were detected, chromium(III) gives a dichloronium ion which is probably tetrasolvated $[Cl_2Cr(NCCH_3)_4]^+$. The following compounds were isolated in these systems[51]:

$V[SbCl_6] \cdot 8\,AN$ (blue), $V_2[SnCl_6]_3 \cdot 12\,AN$ (pink), $[CrCl_2][SbCl_6] \cdot 4\,AN$ (greenish-grey), $[CrCl_2]_2[SnCl_6] \cdot 8\,AN$ (grey), $Cr[SbCl_6]_3 \cdot 6\,AN$ (light brown), $VCl_3 \cdot 3\,AN$ (green), and $CrCl_3 \cdot 3\,AN$.

Titanium salts of the general type $[(C_2H_5)_4N][TiX_4(AN)_2]$ were obtained pure from the reactions of TiX_3 with excess $(C_2H_5)_4NX$ in acetonitrile and were found to behave as 1 : 1 electrolytes in this solvent. Upon thermal decomposition *in vacuo* at 100° they yield salts of empirical formula $[(C_2H_5)_4N][TiX_4]$[52]. For $(PyH)_3TiCl_6$ and $(PyH)_3VCl_6$ the reversible reactions take place:

$$(PyH)_3MCl_6 + 2\,AN \rightleftharpoons PyH[MCl_4(AN)_2] + 2\,PyHCl$$

but $(PyH)_3CrCl_6$ does not dissolve in acetonitrile and this, together with the non-reaction of $CrCl_3Py_3$ and hydrogen chloride is consistent with the inertness of d^3-complexes[52].

---

[48] POPOV, A. I., and E. H. SCHMORR: J. Amer. Chem. Soc. **74**, 4672 (1952). — A. I. POPOV, and F. B. STUTE: J. Amer. Chem. Soc. **78**, 5737 (1956).

[49] POPOV, A. I., and J. N. JESSUP: J. Amer. Chem. Soc. **74**, 6127 (1952).

[50] CLARK, R. J. H., J. LEWIS, D. J. MACHIN, and R. S. NYHOLM: J. Chem. Soc. **1963**, 379.

[51] GUTMANN, V., G. HAMPEL, and W. LUX: Mh. Chem. **96**, 533 (1965).

[52] FOWLES, G. W. A., and B. J. RUSS: J. Chem. Soc. (A), **1967**, 517.

The dichlorides of cobalt, nickel and copper are also capable of providing chloride ions for strong chloride ion acceptors giving hexasolvated transition metal cations and the respective chlorometallate ions[53]:

$$CoCl_2 + 6AN + 2SbCl_5 \rightleftharpoons [Co(AN)_6]^{2+} + 2[SbCl_6]^-; \quad K_7 = 30$$
$$CoCl_2 + 6AN + SnCl_4 \rightleftharpoons [Co(AN)_6]^{2+} + [SnCl_6]^{2-}; \quad K_8 = 3$$
$$CoCl_2 + 6AN + TiCl_4 \rightleftharpoons [Co(AN)_6]^{2+} + [TiCl_6]^{2-}; \quad K_9 = 0.3$$

By all experiments the order of chloride ion acceptor strength

$$SbCl_5 > SnCl_4 > TiCl_4$$

is confirmed.

Colour indicators, such as crystal violet[54], may be used to follow chloride ion transfer reactions, the reactions being analogous to those described in phosphorus oxychloride[55, 56] (p. 124 and 125).

$$CV^+Cl^- + 2SbCl_5 \rightleftharpoons [CV \cdot SbCl_5]^+[SbCl_6]^-$$
pink   green
$$[CV(SbCl_5)]^+ + 2SbCl_5 \rightleftharpoons [CV(SbCl_5)_3]^+$$
green   yellow

Molybdenum(V) chloride, tungsten(VI) chloride, tungsten(V) chloride and tungsten(V) bromide are reduced by acetonitrile to the tetravalent compounds[57] $MX_4(AN)_2$, which are characterized as six coordinate complexes. Likewise vanadium(IV) chloride is reduced to the solvate of the trichloride $VCl_3(AN)_3$[58].

Various other solvate complexes can be obtained, when transition metals are reacted with the free halogens in acetonitrile[59]. In this way $TiCl_4(AN)_2$, $TiBr_4(AN)_2$ and a polyhalide $[Ti(AN)_6][I_3]_3$ are produced. Chlorine, bromine and iodine oxidize vanadium to trivalent compounds, namely $VCl_3(AN)_3AN$, $[VBr_2(AN)_4]Br$ and the polyiodide $[V(AN)_6][I_3]_3$. With chromium $CrCl_3(AN)_3AN$, a bromide of variable composition and $[CrI_2(AN)_4]I$ are obtained[43].

Redox reactions in acetonitrile have been used to prepare a number of coordination compounds of thallium(III), such as[60]:

$$TlCl + Cl_2 + [(C_6H_5)_4As]Cl \rightleftharpoons [(C_6H_5)_4As]TlCl_4$$
$$TlBr + Br_2 + [R_4N]Cl \rightleftharpoons [R_4N]TlBr_4$$

which were found to behave as 1 : 1 electrolytes in solutions of acetonitrile.

Hexachloroprotactinates(IV), as well as the corresponding bromo- and iodo-complexes, have been prepared from the tetrahalides with halide ion donors in acetonitrile[61]. Likewise hexabromotantalates(V) and hexabromoniobates(V) have been obtained in this solvent[62]. The spectra of a variety of complex halides $[MX_6]^{n-}$

[53] GUTMANN, V., G. HAMPEL, and J. R. MASAGUER: Mh. Chem. **94**, 822 (1963).
[54] GUTMANN, V., and K. FENKART, Unpublished.
[55] GUTMANN, V.: Oest. Chem. Ztg. **65**, 273 (1964).
[56] GUTMANN, V., H. HUBACEK, and A. STEININGER: Mh. Chem. **95**, 678 (1964).
[57] ALLEN, E. A., B. J. BRISDON, and G. W. A. FOWLES: J. Chem. Soc. **1964**, 4531.
[58] DUCKWORTH, M. W., G. W. A. FOWLES, and R. A. HODDLESS: J. Chem. Soc. **1963**, 5665.
[59] HATHAWAY, B. J., and D. G. HOLAH: J. Chem. Soc. **1965**, 537.
[60] COTTON, F. A., B. F. G. JOHNSON, and R. M. WING: Inorg. Chem. **4**, 502 (1965).
[61] BROWN, D., and P. J. JONES: J. Chem. Soc. (A), **1967**, 243.
[62] BROWN, D., and P. J. JONES: J. Chem. Soc. (A), **1967**, 247.

where $M = U(IV)$, $Np(IV)$, $Pu(IV)$, $W(V)$ and $X = Cl$, Br, I have been measured in acetonitrile and there is no evidence for reaction with the solvent[63, 64].

Transition metal cations are usually octahedrally solvent coordinated according to spectral evidence. When competitive ligands, such as bromide, chloride or azide ions are added to the solutions of transition metal perchlorates or tetra-fluoroborates, replacement of the solvent molecules together with the coordination changes involved has been followed by spectrophotometric, conductometric and potentiometric techniques[3] (Table 67).

It has been shown that in the nickel(II) system even a weakly coordinating ligand, such as the perchlorate ion, may approach the coordination sphere of the metal ion in acetonitrile[65]; the following hexacoordinated compounds were prepared and characterized:

$$[Ni(AN)_6][ClO_4]_2, \quad [Ni(AN)_4(ClO_4)_2] \quad \text{and} \quad [Ni(AN)_2(ClO_4)_2]$$

In the first compound the perchlorate ion is not coordinated to the nickel ion, but it acts as a monodentate ligand in the second and as a bidentate ligand in the third compound[65].

The stepwise formation of bromide coordinated species of cobalt(II) has been observed; $[BrCo(AN)_5]^+$ is apparently octahedral while $[CoBr_2(AN)_2]$, $[CoBr_3AN]^-$ and $[CoBr_4]^{2-}$ are tetrahedral[66, 67]. In the chloride system of cobalt(II) no cationic chloro-complex has been found, but the species $[CoCl_2]$, $[CoCl_3]^-$ and $[CoCl_4]^{2-}$ are produced which seem to have tetrahedral arrangements[53, 68, 69].

The presence of only two species is indicated in the cobalt(II) azidosystem; these are namely $Co(N_3)_2(AN)_2$, with apparently distorted tetrahedral configuration and the tetraazido-complex anion. The distorted structure of the solvated diazide may be responsible for the absence of the formation of the triazido anionic complex[70].

In the corresponding nickel system all bromide-coordinated species seem to be formed[66], but in the chloride system only $[NiCl_3AN]^-$ and $[NiCl_4]^{--}$ have been detected. (Nickel(II)chloride has a low solubility in acetonitrile.) With azide two coordination forms are probably produced, which are hexacoordinated and of the general formula[70] $[Ni(N_3)_a(AN)_{6-a}]^{+2-a}$.

In the copper(II) bromide system $CuBr^+$ and $CuBr_2$ are unstable[73-75] since they are decomposed to copper(I) compounds and free bromine. $[CuBr_3AN]^-$ appears to be more stable than $[CuBr_4]^{--}$, but in the chloro-system $[CuCl_4]^{--}$ has a high stability, as also has $[CuCl_3(AN)]^-$. The stability constants in AN ($\log K = 9.7$ for $[CuCl]^+$, 7.9 for $CuCl_2$, 7.1 for $[CuCl_3]^-$ and 3.7 for $[CuCl_4]^{--}$)

[63] RYAN, R. L., and C. K. JØRGENSEN: Mol. Phys. 7, 17 (1964).

[64] BRISDON, B. J., and R. A. WALTON: J. Chem. Soc. 1965, 2274.

[65] WICKENDEN, A. E., and R. A. KRAUSE: Inorg. Chem. 4, 404 (1965).

[66] GUTMANN, V., and K. FENKART: Mh. Chem. 98, 1 (1967).

[67] JANZ, G. J., A. E. MARCINKOVSKY, and H. V. VENKATASETTY: Electrochim. Acta 8, 867 (1963).

[68] BAAZ, M., V. GUTMANN, G. HAMPEL, and J. R. MASAGUER: Mh. Chem. 93, 1416 (1962).

[69] LIBUS, W.: Proc. 7. ICCC, Stockholm 349 (1962).

[70] GUTMANN, V., and O. LEITMANN: Mh. Chem. 97, 926 (1966).

[71] GUTMANN, V., and O. BOHUNOVSKY: Mh. Chem. in press (1968).

[72] GUTMANN, V., and H. BARDY: Unpublished.

[73] SCHNEIDER, W., and A. V. ZELEWSKY: Helv. Chim. Acta 46, 1848 (1963).

are higher by several orders of magnitude than in water[74-76]. The solutions show moderate conductivities indicating some autocomplex formation of cupric chloride. Structural investigations have been carried out on the compounds $Cu_2Cl_4(AN)_2$ and $Cu_3Cl_6(AN)_3$ which were shown to be essentially planar with the

Table 67. *Coordination Forms Involving Transition Metal Ions in Solutions of Acetonitrile*

System / Competitive Ligands coordinated	0	1	2	3	4	5	6	References
$Co^{++} + Br^-$	+	+	+	+	+			66, 67
$Co^{++} + Cl^-$	+	−	+	+	+			53, 60, 67–69
$Co^{++} + N_3^-$	+	−	+	−	+			70
$Co^{++} + I^-$	+		+	+				71
$Co^{++} + NCS^-$	+	+	−	+	+			71
$Ni^{++} + I^-$	+	+	+	+	+			72
$Ni^{++} + Br^-$	+	+	+	+	+			66
$Ni^{++} + Cl^-$	+	−	↓	+	+			53, 68
$Ni^{++} + N_3^-$	+	−	+	−	+		?	72
$Ni^{++} + CNS^-$	+	+		+	+		+	72
$Cu^{++} + Br^-$	+	−	−	+	+			42, 73–75
$Cu^{++} + Cl^-$	+	?	?	+	+			53, 68, 74, 76
$Cu^{++} + N_3^-$	+	+	+	+	+			70
$Mn^{++} + Br^-$	+	+	+	+	+			77
$Mn^{++} + Cl^-$	+	+	+	?	+			78
$Mn^{++} + N_3^-$	+	−	+	−				79
$VO^{++} + Br^-$	+	+	+		+			80 a
$VO^{++} + Cl^-$	+	+	+	−	+			80
$VO^{++} + N_3^-$	+	+	+	−	−	+		80
$VO^{++} + NCS^-$	+	+	+	+				80 a
$Ti^{3+} + Cl^-$	+	−	−	+	+	−	−	81
$Ti^{3+} + N_3^-$	+	−	−	+	+	−	+	82
$V^{3+} + Br^-$	+	−	+	+	+	−	−	77
$V^{3+} + Cl^-$	+	−	−	+	+	−	−	81
$V^{3+} + N_3^-$	+	+	−	+	−	−	+	82
$Cr^{3+} + Cl^-$	+	−	+	+	+	−	+	81
$Cr^{3+} + N_3^-$	+	−	+	+	−	−	+	82
$Fe^{3+} + Cl^-$	+	+	−	+	+	−	−	78
$Fe^{3+} + N_3^-$	+	−	−	+	−	−	−	79

[74] BARNES, J. C., and D. N. HUME: Inorg. Chem. **2**, 444 (1963).

[75] FURLANI, C., A. SGAMELLATTI, and G. CIULLO: Ric. Sci., Rend. Sez. A **4**, 49 (1964); C. A. **61**, 1404 f. (1964).

[76] MANAHAN, S. E., and R. T. IWAMOTO: Inorg. Chem. **4**, 1409 (1965).

[77] GUTMANN, V., and K. FENKART: Mh. Chem. **98**, 286 (1967).

[78] GUTMANN, V., and W. K. LUX: Mh. Chem. **98**, 276 (1967).

[79] GUTMANN, V., and W. K. LUX: J. Inorg. Nucl. Chem., **29**, 2391 (1967).

[80] GUTMANN, V., and H. LAUSSEGGER: Mh. Chem. **98**, 439 (1967).

[80a] GUTMANN, V., and H. LAUSSEGGER: Mh. Chem. in press (1968).

[81] GUTMANN, V., A. SCHERHAUFER, and H. CZUBA: Mh. Chem. **98**, 619 (1967).

[82] GUTMANN, V., O. LEITMANN, A. SCHERHAUFER, and H. CZUBA: Mh. Chem. **98**, 188 (1967).

acetonitrile molecules in *trans*-positions and with chlorine bridging[83]. All possible coordination forms have been detected in the copper(II) azidosystem.

In the manganese-bromide system, the octahedral $[Mn(AN)_6]^{++}$ ion is converted to a tetrahedral species[77] by coordination of one bromide ligand to give $[MnBr(AN)_3]^+$. The other tetrahedral forms, $[MnBr_2(AN)_2]$, $[MnBr_3AN]$ and $[MnBr_4]^{--}$ can be obtained by further addition of bromide ions to the acetonitrile solution[77,84]. With chloride ions also tetrahedral species are produced but $[MnCl_3AN]^-$ has a low stability in acetonitrile.

Species with octahedral or rather distorted octahedral structures are formed in the VO^{++}, Ti^{3+}, Cr^{3+} and V^{3+} systems with chloride and azide ions[3,80-82].

Iodides of Mn(II), Co(II) and Ni(II) seem to undergo autocomplex formation to give $[M(AN)_4][MI_4]$[71,72,85].

Ferric ion was found to coordinate chloride ions fairly easily to give tetrahedral species[78]:

$$[Fe(AN)_6]^{3+} + Cl^- \rightleftharpoons [FeCl(AN)_3]^{2+} + 3\,AN$$
$$[FeCl(AN)_3]^{2+} + 2\,Cl^- \rightleftharpoons [FeCl_3AN] + 2\,AN$$
$$[FeCl_3AN] + Cl^- \rightleftharpoons [FeCl_4]^- + AN$$

but with azide ions the octahedral species appear to be formed. The mono-azido complex, known to occur in aqueous solutions, has not been detected in acetonitrile[79].

Conductance studies indicate the formation of $[HgX_3]^-$, $[HgX_4]^{--}$ and $[Hg_2X_5]^-$ when X^--ligands are added to HgX_2-solutions in acetonitrile[86].

While alkali metals and alkaline earth metals cannot be obtained by electrolysing the appropriate acetonitrile solutions[87], other metals have been electrodeposited. The following salts are electrolytically reduced to the respective metals[88]: Cu_2Cl_2, Cu_2Br_2, $AgNO_3$, MnI_2, $CoCl_2$, $NiBr_2$, $ZnCl_2$, CdI_2, $SnCl_2$, $AsCl_3$ and $SbCl_3$. When solutions of silver nitrate are electrolysed with the anodic materials consisting of Co, Ni, Zn or Cd the following compounds were obtained at the anodes respectively $Co(NO_3)_2(AN)_3$, $Ni(NO_3)_2(AN)_2$, $Zn(NO_3)_2(AN)_2$ and $Cd(NO_3)_2(AN)_2$. With solutions of silver perchlorate the compounds $Sb(ClO_4)_3$, $Bi(ClO_4)_3$ and $Sn(ClO_4)_2(AN)_2$ were produced in an analogous manner. All attempts to obtain titanium and other refractory metals by electrolysis from acetonitrile solutions were unsuccessful[89].

Acetonitrile is a useful medium for many other types of complex reactions. It was found to be coordinated in certain carbonyl compounds[90-93], such as $W(CO)_4(AN)_2$ or $W(CO)_3(AN)_3$ in which the acetonitrile molecules are attached

[83] WILLETT, R. D., and R. E. RUNDLE: J. Chem. Phys. **40**, 838 (1964).

[84] COTTON, F. A., D. M. L. GOODGAME, and M. GOODGAME: J. Amer. Chem. Soc. **84**, 167 (1962).

[85] HATHAWAY, B. J., and D. G. HOLAH: J. Chem. Soc. **1964**, 2400.

[86] ELLENDT, G., and K. CRUSE: J. Phys. Chem. **201**, 130 (1952).

[87] MÜLLER, R., E. PINTER, and K. PRETT: Mh. Chem. **45**, 525 (1925).

[88] SCHMIDT, H.: Z. anorg. allg. Chem. **271**, 305 (1953).

[89] GUTMANN, V., and G. SCHÖBER: Oest. Chem. Ztg. **59**, 321 (1958).

[90] STROHMEIER, W., and K. GERLACH: Z. Naturf. **15 b**, 622 (1960).

[91] STROHMEIER, W., and G. SCHÖNAUER: Chem. Ber. **94**, 1346 (1961).

[92] DOBSON, G. R., M. F. AMR EL SAYED, I. W. STOLZ, and R. K. SHELINE: Inorg. Chem. **1**, 526 (1962).

[93] TATE, D. B., W. R. KNIPPLE, and J. M. AUGL: Inorg. Chem. **1**, 433 (1962).

to the metal atoms through the N-atom[94]. These complexes are excellent intermediates in the formation of new compounds not available by other routes[93].
Sulphur(IV) fluoride is known to be readily produced in acetonitrile solutions:

$$4\,NaF + 2\,S_2Cl_2 \rightarrow SF_4 + 3\,S + 4\,NaCl$$

5. Sulpholane (Tetramethylenesulphone) ($DN_{SbCl_5} = 14.8$)

Table 68. *Some Physical Properties of Sulpholane*

Melting Point (°C)	+27.5
Boiling Point (°C)	+285
Density at 30° ($g \cdot cm^{-3}$)	1.265
Dielectric Constant at 30°	42
Donor Number DN_{SbCl_5}	14.8

Sulpholane is a solvent of moderate donor properties with a donor number
similar to that of acetonitrile, but its capacity for the solvation of metal ions[95]
seems to be different from that of acetonitrile. It forms an adduct with boron(III)
fluoride, but no compound has been isolated with phosphorus(V) fluoride[96]. It was
found to react also with phenol with $\Delta H = -4.9\,kcal \cdot mole^{-1}$ and in the result-
ant compound coordination through hydrogen bridges has been assumed[97].

Sulpholane appears to act as a bidentate ligand towards metal ions, such as
cobalt(II). A pink complex of composition $[CoD_3][ClO_4]_2$ (D = sulpholane) has
been described, which is unstable towards water[98]. Assuming a coordination number
of six, in pink Co(II)-complexes two coordination sites ought to be occupied by
one solvent molecule. The same conclusion is drawn from the blue compound
of composition $CoDCl_2$, which according to its spectrum with peaks at 687 and
590 nm and $\log \varepsilon \sim 2.5$ should be tetrahedral.

Some conductivity measurements have been reported[99]. Nitrations have been
carried out with nitrylfluoroborate in sulpholane[100-102]. Derivatives of $[B_{10}H_{10}]^{2-}$
and $[B_{12}H_{12}]^{2-}$ with various Lewis bases were obtained in this solvent[102], which

[94] STOLZ, I. W., R. G. DOBSON, and R. K. SHELINE: Inorg. Chem. **2**, 322 (1963).
[95] ARNETT, M. E., and C. F. DOUTY: J. Amer. Chem. Soc. **86**, 409 (1964).
[96] JONES, J. G.: Inorg. Chem. **5**, 1229 (1966).
[97] DRAGO, R. S., B. WAYLAND, and R. L. CARLSON: J. Amer. Chem. Soc. **85**, 3125 (1963).
[98] LANGFORD, C. H., and P. O. LANGFORD: Inorg. Chem. **1**, 184 (1962).
[99] FERNANDEZ-PRINI, R., and J. E. PRUE: Trans. Farad. Soc. **62**, 1257 (1966).
[100] OLAH, G. A., S. J. KÜHN, and S. H. FLOOD: J. Amer. Chem. Soc. **83**, 4571 (1961).
[101] KREIENBUEHL, P., and H. ZOLLINGER: Tetrahedron Letters **1965**, 1739.
[102] MILLER, H. C., W. R. HERTLER, E. L. MUETTERTIES, W. H. KNOTH, and N. E. MILLER:
Inorg. Chem. **4**, 1216 (1965).

has also been used in the study of substitution reactions in coordination compounds[103]. Sulpholane has also been used for liquid-liquid extractions, and offers certainly many possibilities as a reaction medium for coordination chemistry.

6. Propanediol-1,2-carbonate (PDC) $(DN_{SbCl_5} = 15.1)$

Propanediol-1,2-carbonate is a very useful solvent of medium donor properties, similar to those of sulpholane or acetonitrile. It has a high dielectric constant so that dissolved compounds, which are ionized, will be extensively electrolytically dissociated. Its stability towards redox reactions is another useful property, and furthermore it is easy to purify and to handle.

While the acetonitrile molecule is small and rod-like in shape the propanediol-1,2-carbonate molecule is large and bulky, and this may account for a number of differences[3] observed in the two solvents of similar donor number.

Table 69. *Physical Properties of Propanediol-1,2-carbonate*

Melting Point (°C)	−49.2
Boiling Point (°C)	241.7
Density at 20° (g · cm^{-3})	1.2057
Viscosity at 38° (Centipoise)	2.013
Dielectric Constant at 20°	70
Specific Conductivity at 20° (Ohm^{-1} · cm^{-1})	10^{-8}
Dipole Moment (Debye)	5.2
Donor Number DN$_{SbCl_5}$	15.1
Enthalpy of Vaporization (kcal · mole^{-1})	11.90
Trouton-Constant (cal · mole^{-1} · deg^{-1})	23.19

Coordination chemistry involving acceptor molecules is expected to be similar to that in acetonitrile and, in fact, many solvent-acceptor compounds may be isolated from its solutions. Such solvates are likely to be converted into the respective complex compounds if competitive ligands are added.

With transition metal ions hexacoordinated species are formed by interactions with the solvent molecules. Since the PDC-molecule is rather large, most of the transition metal ions will be completely shielded by the coordination shell consisting usually of six solvent molecules. Ligand exchange reactions are therefore less easily accomplished than with the corresponding species in acetonitrile. Due to the steric properties of the solvent molecules the tendency to form species with different ligands is again much less pronounced than in acetonitrile and the number of coordination forms obtained in propanediol-1,2-carbonate solutions is small. High stabilities are, however, found for symmetrical coordination forms. Thus the steric properties of the solvent molecules seem to be a decisive factor both in the formation and for the stabilities of the coordination forms in its solutions[3].

The dibromides of cobalt(II), nickel(II) and manganese(II), vanadium(III) bromide, as well as the diiodides of cobalt (II) and nickel (II) are completely

---

[103] HUGHES, M. E., and M. L. TOBE: J. Chem. Soc. **1965**, 1204.

subject to autocomplex formation[65, 66, 77]. When bromide ions are added to the solvated metal cations only one coordination change is observed in each system to give the highly symmetrical bromometallate, which is usually tetrahedral in structure. No mixed species, as observed in acetonitrile, can be formed in the bromo systems under consideration in propanediol-1,2-carbonate.

Table 70. *Coordination Forms Found in Solutions of Propanediol-1,2-carbonate*

System	Competitive Anions coordinated						References
	1	2	3	4	5	6	
$Co^{++} + I^-$	−	−	−	+			71
$Co^{++} + Br^-$	−	−	−	+			66
$Co^{++} + Cl^-$		+		+			71
$Co^{++} + NCS^-$	+	+	+	+			71
$Co^{++} + CN^-$	−	+	−	−	+		71
$Ni^{++} + I^-$	−	−	−	+			72
$Ni^{++} + Br^-$	−	−−	−	+			66
$Ni^{++} + Cl^-$		+		+			72
$Ni^{++} + N_3^-$	−	↓	−	+			72
$Ni^{++} + NCS^-$	−	↓	−	+	−	+	72
$Ni^{++} + CN^-$	−	↓	−	+			72
$Mn^{++} + Br^-$	−	−	−	+			77
$Mn^{++} + Cl^-$	−	+	−	−	−	+	78
$Mn^{++} + N_3^-$	−	+	−	+	−	−	79
$VO^{++} + Br^-$	+	+	−	+			80 a
$VO^{++} + Cl^-$	+	+	+	+	−	−	80
$VO^{++} + N_3^-$		+	−	−	+	−	80
$VO^{++} + NCS^-$	+	+	+	+			80 a
$Ti^{3+} + Cl^-$	+	−	+	+	−	?	81
$Ti^{3+} + N_3^-$	−	−	+	−	−	+	82
$V^{3+} + Br^-$	−	−	−	+			77
$V^{3+} + Cl^-$	+	−	+	+	−	−	81
$V^{3+} + N_3^-$	+	−	+	+	−	+	82
$Cr^{3+} + Cl^-$	−	+	+	+	−	−	81
$Cr^{3+} + N_3^-$	−	+	+	−	−	+	82
$Fe^{3+} + Cl^-$	+	−	+	+	−	−	78
$Fe^{3+} + N_3^-$	+	+	+	−	−	+	79

↓ = insoluble

A further contribution in assisting autocomplex formation is provided by the high dielectric constant of the solvent.

$$2\,CoBr_2 + 6\,PDC \rightleftharpoons [Co(PDC)_6]^{2+} + [CoBr_4]^{--}$$
$$\text{octahedral} \qquad \text{tetrahedral}$$
$$4\,VBr_3 + 6\,PDC \rightleftharpoons [V(PDC)_6]^{3+} + 3\,[VBr_4]^{-}$$
$$\text{octahedral} \qquad \text{tetrahedral}$$

In the manganese-chloride system $MnCl_2(AN)_4$ is formed[78] which may partially give ions due to autocomplex formation. Again no highly unsymmetrical forms have been observed in PDC solutions and it seems likely that $[MnCl_6]^{4-}$ is the only anionic chloro-complex formed in this medium. It is remarkable that in the manganese(II) azide system $[Mn(N_3)_4(PDC)_2]^{2-}$ seems to be the stable anionic species[79].

In the vanadyl systems the species existing in PDC-solutions appear to have distorted octahedral structures[80]. The highest chloride-coordinated ion is $[VOCl_4]^{2-}$, which may be present to some extent in the absence of excess chloride ions, due to some degree of autocomplex formation by vanadium oxydichloride[80]. Likewise the ultimate azidocomplex seems to have the composition $[VO(N_3)_4]^{2-}$. It has been concluded from the spectra that all vanadyl species in propanediol-1,2-carbonate are of lower symmetries than the corresponding species in acetonitrile or in dimethyl sulphoxide, and this has been ascribed to stronger π-bonding contributions to the $V=O$ bond in propanediol-1,2-carbonate[80].

The existence of the hexachlorotitanate ion in propanediol-1,2-carbonate is doubtful[81] but the hexaazido-complex is readily formed[82].

$$[Ti(PDC)_6]^{3+} + 3\,N_3^- \rightleftharpoons [Ti(N_3)_3(PDC)_3] + 3\,PDC$$
$$[Ti(N_3)_3(PDC)_3] + 3\,N_3^- \rightleftharpoons [Ti(N_3)_6]^{3-} + 3\,PDC$$

With vanadium(III) and chromium(III) the tetrachloro-complexes and the hexa-azidometalles are formed[81,82].

The ferric-systems in propanediol-1,2-carbonate are similar to those in acetonitrile[78]: tetrahedral $[FeCl]^{++}$, $FeCl_3$ and $[FeCl_4]^-$ are observed in each case:

$$[Fe(PDC)_6]^{3+} + Cl^- \rightleftharpoons [FeCl(PDC)_3]^{2+} + 3\,PDC$$
$$\text{octahedral} \qquad\qquad\qquad \text{tetrahedral}$$
$$[FeCl(PDC)_3]^{2+} + 2\,Cl^- \rightleftharpoons [FeCl_3(PDC)] + 2\,PDC$$
$$[FeCl_3(PDC)] + Cl^- \rightleftharpoons [FeCl_4]^- + PDC$$

In the iron(III)-azido system pseudooctahedral cationic species are formed such as $[FeN_3]^{2+}$, which were not detected in acetonitrile, but are known to exist in aqueous solutions; the highest azide coordinated species is the hexaazido-ferrate(III)[79].

It is to be expected that extensive use will be made of this solvent both as a medium to carry out reactions and for many electrochemical and redox-studies.

7. Acetone (DNSbCl₅ = 17.0)

Acetone is a very useful solvent, but it is extremely difficult to remove the last traces of water. Consequently nearly all of the investigations have been carried out with acetone containing certain amounts of water. The donor number is nearly as high as that of water and it is apparent that the possibilities of applying anhydrous acetone as a reaction medium in coordination chemistry have scarcely been explored.

Table 71. *Physical Properties of Acetone*

Melting Point (°C)	−95.4
Boiling Point (°C)	+56.2
Density at 25° (g · cm⁻³)	0.785
Viscosity at 30° (Centipoise)	0.295
Dielectric Constant at 25°	20.7
Specific Conductivity at 25° (Ohm⁻¹ · cm⁻¹)	$6 \cdot 10^{-8}$
Dipole Moment (Debye)	2.84 (gas)
Donor Number DN_{SbCl_5}	17.0
Enthalpy of Vaporization (kcal · mole⁻¹)	6.953
Trouton-Constant (cal · mole⁻¹ · deg⁻¹)	21.12

Table 72. *Some Adducts of Acetone with Acceptor Halides*

Acceptor	Number of Solvent Molecules Coordinated	$-\Delta H$	References
BBr₃	1	0.7	104
GaCl₃	1	15.3	104
SbCl₅	1	17.0	5, 6
AlBr₃	2		105
SnCl₄	2		105

Spectrophotometric studies have been made on the ligand exchange between the hydrated nickel(II) ion and bromide ions[106]. In those systems which were, however, not anhydrous the species $[NiBr]^+$, $NiBr_2$, $[NiBr_3]^-$ and $[NiBr_4]^{--}$ were detected; nickel(II) bromide was observed to have a low stability, thus indicating considerable autocomplex formation[106,107].

$$NiBr_2 \rightleftharpoons NiBr^+ + NiBr_3^-$$

The cobalt(II) systems with chloride, bromide and iodide ions as ligands have also been investigated by the spectrophotometric method[13,108]. The coordination forms CoX_2, $[CoX_3]^-$ and $[CoX_4]^{--}$ were detected. $[CoCl_4]^{--}$ is largely dissociated but $[CoBr_4]^{--}$ and $[CoI_4]^{--}$ are dissociated to an even greater extent. This is exemplified by values determined for K in the expression:

$$K = \frac{[CoX_4]^{--}}{[CoX_3]^- \cdot X^-}$$

and it has been shown that for $X = Cl^-$ $(K \sim 5 \cdot 10^{-2})$, $X = Br^-$ $(K \sim 42)$ and $X = I^-$ $(K \sim 16)$. The spectra of the corresponding complexes of the different halides indicated an increase in ligand field strengths in the order $I^- < Br^- < Cl^-$ in agreement with the spectrochemical series.

Acetone is a useful medium to prepare iodothallates[60] or azido-complexes of various elements[109].

$$[(C_6H_5)_4As][TlCl_4] + 4\,NaI \rightleftharpoons [(C_6H_5)_4As][TlI_4] + 4\,NaCl$$
$$[(C_2H_5)_4N]_2[MnBr_4] + 4\,AgN_3 \rightleftharpoons [(C_2H_5)_4N]_2[Mn(N_3)_4] + 4\,AgBr$$

[104] GREENWOOD, N. N., and P. G. PERKINS: J. Chem. Soc. **1960**, 356.

[105] GORENBEIN, E. YA., and V. V. SUKHAN: Zh. Neorg. Khim. 8, 360 (1963); C. A. **58**, 13405 (1963).   [106] FINE, D. A.: Inorg. Chem. 4, 345 (1965).

[107] SRAMKO, T.: Chem. Zvesti **17**, 725 (1963); C. A. **60**, 8697 b (1964).

[108] FINE, D. A.: J. Amer. Chem. Soc. **82**, 1139 (1962).

[109] BECK, W., K. FELDL, and E. SCHUIERER: Angew. Chem. **77**, 458 (1965); Int. Ed. **4**, 439 (1965).

Molybdenum(V) chloride was found to react with ammonium thiocyanate according to the equation[110]:

$$(CH_3)_2CO + MoCl_5 + 4\,NH_4CNS \rightleftharpoons 4\,NH_4Cl + [MoCl(NCS)_4OC(CH_3)_2]$$

8. Ethyl Acetate ($DN_{SbCl_5} = 17.1$)

Ethyl acetate has a donor number comparable to that of water[1], but it has less favourable steric properties. Thus the stabilities of highly solvent-coordinated species are lower than in water. Another disadvantage is the low dielectric constant.

Table 73. *Physical Properties of Ethyl Acetate*

Melting Point (°C)	−85.0
Boiling Point (°C)	+77.1
Density at 25° (g · cm^{-3})	0.89455
Viscosity (Centipoise) at 25°	0.426
Dielectric Constant at 25°	6.02
Specific Conductivity at 25° (Ohm^{-1} · cm^{-1})	$3 \cdot 10^{-9}$
Dipole Moment (Debye)	1.76 (gas)
Donor Number DN_{SbCl_5}	17.1
Enthalpy of Vaporization (kcal · mole^{-1})	7.710
Trouton-Constant (cal · mole^{-1} · deg^{-1})	22.02

Many crystalline donor-acceptor compounds have been found, such as 1 : 1 compounds with $TeCl_4$, SO_3, $SbCl_3$, $TiCl_4$, $AlCl_3$, $SbCl_5$, $FeCl_3$, BF_3, BCl_3, $FeCl_2$, $MgCl_2$ and $SnCl_4$ and disolvates of $FeCl_3$, SO_3, $SnCl_4$, $SnBr_4$, $ZrCl_4$,$ZrBr_4$ and ZrI_4.

Infrared studies have shown that the carbonyl oxygen atom of methyl acetate functions as the donor atom[111]. The compound with boron(III) fluoride has been investigated in the molten state and was found to be oxygen-coordinated, but considerably ionized[112].

PAUL and coworkers[113,114] assumed the existence of a self-ionization in the pure solvent according to the equation:

$$CH_3COOC_2H_5 \rightleftharpoons [C_2H_5]^+ + [CH_3COO]^-$$

This would involve splitting of a carbon-oxygen bond and require the stability of the ethyl-ion. Various nitrogenous bases give conducting solutions. Fluorosulphonic acid and sulphuric acid are good electrolytes, while trichloroacetic and acetic acids are very poor conductors in ethyl acetate solutions.

$SnCl_4$, $TiCl_4$, SO_3 and $SbCl_5$ were found to give conducting solutions possibly due to autocomplex formation equilibria, in ethyl acetate (D), such as

$$2\,D \cdot SbCl_5 \rightleftharpoons [D_2SbCl_4]^+ + [SbCl_6]^-$$

Metals such as iron are oxidized by the free halogens[115]:

$$2\,Fe + 2\,Br_2 + n\,CH_3COOC_2H_5 = [Fe^{II}(CH_3COOC_2H_5)_n][FeBr_4]_2$$

[110] FUNK, H., and H. BOEHLAND: Z. anorg. allg. Chem. **324**, 168 (1963).
[111] ZACKRISSON, M., and I. LINDQVIST: J. Inorg. Nucl. Chem. **17**, 69 (1961).
[112] GREENWOOD, N. N., and R. L. MARTIN: J. Chem. Soc. **1953**, 751.
[113] PAUL, R. C., and K. C. MALHOTRA: Z. anorg. allg. Chem. **321**, 56 (1963).
[114] PAUL, R. C., D. SINGH, and K. C. MALHOTRA: Z. anorg. allg. Chem. **321**, 70 (1963).
[115] HATHAWAY, B. J., and D. G. HOLAH: J. Chem. Soc. **1964**, 2408.

9. Diethylether $(DN_{SbCl_5} = 19.2)$

Diethylether is used as a solvent in a variety of reactions in organic chemistry and more recently it has been found an extremely useful medium in hydride chemistry.

The donor properties of diethylether are similar to those of water, but water is, of course, the more powerful solvent. Diethylether differs from water by its unfavourable steric properties, by its low dielectric constant and by its poor association in the liquid state.

Table 74. *Some Physical Properties of Diethylether*

Melting Point (°C)	−116.4
Boiling Point (°C)	+34.6
Density at 25° $(g \cdot cm^{-3})$	0.70778
Viscosity at 20° (Centipoise)	0.242
Dielectric Constant at 20°	4.4
Specific Conductivity at 20° $(Ohm^{-1} \cdot cm^{-1})$	10^{-12}
Dipole Moment (Debye)	1.2 (in benzene)
Donor Number DN_{SbCl_5}	19.2
Enthalpy of Vaporization $(kcal \cdot mole^{-1})$	6.357
Trouton-Constant $(cal \cdot mole^{-1} \cdot deg^{-1})$	20.66

Numerous donor-acceptor compounds are known in which diethylether acts as the donor molecule and considerable heat is involved in their formation[1,116]. Examples are $GaCl_3 \cdot O(C_2H_5)_2$[116], $AlCl_3 \cdot O(C_2H_5)_2$ and $AlCl_3[O(C_2H_5)_2]_6$ (stable only at low temperatures)[117], $NbCl_5 \cdot O(C_2H_5)_2$, $TaCl_5 \cdot O(C_2H_5)_2$[118], $BX_3 \cdot O(C_2H_5)_2$[119,120], $[BX_2(O(C_2H_5)_2)_2][BX_4]$ and $[BX_2(O(C_2H_5)_2)_2]X$[120].

Ionization is believed to occur according to the general equation:

$$2 X_nMO(C_2H_5)_2 \rightleftharpoons [X_{n-1}M(O(C_2H_5)_2)_2]^+ + [MX_{n+1}]^-$$

but JANDER and KRAFFCZYK[121] preferred an interpretation which would involve splitting of the solvent molecule:

$$AlBr_3 + (C_2H_5)_2O \rightleftharpoons [C_2H_5]^+ + [AlBr_3OC_2H_5]^-$$

Indeed the postulated self-ionization[121–123]

$$(C_2H_5)_2O \rightleftharpoons [C_2H_5]^+ + [OC_2H_5]^-$$

is unlikely to be of any significance.

[116] GREENWOOD, N. N., and P. G. PERKINS: J. Inorg. Nucl. Chem. **4**, 291 (1957).

[117] TUROVA, YA., N. S. KEDROVA, K. N. SEMENENKO, and A. V. NOVOSELOVA: Zh. Neorgan. Khim. **9**, 905 (1964); C. A. **61**, 1315 (1964).

[118] COPLEY, D. B., F. FAIRBROTHER, and A. THOMPSON: J. Chem. Soc. **1964**, 315.

[119] RUTENBERG, A. C., and A. A. PALKO: J. Chem. Phys. **69**, 527 (1965).

[120] SHCHEGOLEVA, T. H., V. D. SHELUDYAKOV, and B. M. MIKHAILOV: Dokl. Akad. Nauk SSSR **152**, 888 (1963); C. A. **60**, 6455 c (1964).

[121] JANDER, G., and K. KRAFFCZYK: Z. anorg. allg. Chem. **282**, 121 (1955).

[122] JANDER, G., and K. KRAFFCZYK: Z. anorg. allg. Chem. **282**, 217 (1955).

[123] JANDER, G., and H. KNAUER: Z. anorg. allg. Chem. **287**, 138 (1956).

Many ionic compounds are soluble such as bromides and iodides of the more electropositive elements, as well as certain transition metal halides, such as the anhydrous chlorides of nickel(II) and cobalt(II) [29]. The solubility of nickel(II) chloride is decreased with decreasing donor number of the solvent and the following order is found:

$$DMSO > TMP > (C_2H_5)_2O > AN > CH_3NO_2 > C_2H_4Cl_2$$

In the system
$BeCl_2(O(C_2H_5)_2)_2$—$AlCl_3 \cdot O(C_2H_5)_2$ the compound $BeCl_2 \cdot AlCl_3(O(C_2H_5)_2)_3$
has been detected[117].

An interesting example of the influence of the donor number on the mode of ionization of a dissolved compound is the behaviour of boron(III) chloride. It reacts both in diethyl ether and in tetrahydrofurane as a chloride ion donor towards ferric chloride[124]:

$$BCl_3 + FeCl_3 + 2(C_2H_5)_2O \rightleftharpoons [Cl_2B(O(C_2H_5)_2)_2]^+[FeCl_4]^-$$

This behaviour is not found in phosphorus oxychloride[125] (donor number 11.7) but is to a certain extent in phenylphosphonic dichloride[126] ($DN_{SbCl_5} = 18.5$). Thus the formation of a $[BCl_2]^+$ unit is stabilized by solvent-coordination and appears to occur only in solvents of medium or high donor number.

Ionic reactions described by JANDER and coworkers[121–123] lead frequently to insoluble products and may simply be represented as metathetical reactions by the overall equations:

$$AlBr_3 + 3\,TlOC_2H_5 \rightarrow 3\,TlBr \downarrow + Al(OC_2H_5)_3$$
$$Al(ClO_4)_3 + 3\,TlOC_2H_5 \rightarrow 3\,TlClO_4 \downarrow + Al(OC_2H_5)_3$$
$$AgClO_4 + TlOC_2H_5 \rightarrow TlClO_4 \downarrow + AgOC_2H_5$$
$$BBr_3 + 3\,NaOC_2H_5 \rightarrow 3\,NaBr \downarrow + B(OC_2H_5)_3$$

It appears that complex chemistry in this solvent has not received much attention, although numerous reactions should be possible.

Dioxane and tetrahydrofurane are similar to diethylether in donor properties, and as non-aqueous solvents are frequently preferred to diethyl ether.

10. Trimethyl Phosphate (TMP) ($DN_{SbCl_5} = 23$)

Trialkyl phosphates are powerful ionizing solvents, which give stable hexa-coordinated species with most transition metal cations. The dielectric constant of TMP allows reasonable electrolytic dissociation in its solutions.

By virtue of the high donor number of the solvent ligand exchange reactions involving replacement of solvent molecules in the coordination shell by weak competitive ligands are unlikely. Manganese(II) bromide[77] is ionized and no

[124] MIKHAILOV, B. M., T. A. SHCHEGOLEVA, and V. D. SHELUDYAKOV: Izv. Akad. Nauk SSSR Ser. Khim. **1964**, 2165; C. A. **62**, 8988 (1965).

[125] BAAZ, M., V. GUTMANN, and L. HÜBNER: Mh. Chem. **91**, 694 (1960).

[126] BAAZ, M., V. GUTMANN, M. Y. A. TALAAT, and T. S. WEST: Mh. Chem. **92**, 150 (1961).

bromide-coordinated species is found in this solvent even in the presence of excess bromide ions. With vanadium(III) bromide ionization takes place and the co-ordination species $[V(TMP)_6]^{3+}$, $[VBr(TMP)_5]^{2+}$ and $VBr_3(TMP)_3$ have been detected in the solution[77]. The formation of a bromovanadate(III) has not been observed. In the bromo-systems of cobalt(II) and nickel(II) MBr_2 and $[MBr_4]^{--}$ were found[66]. Spectral evidence indicates that tetrahedral units are formed which are slowly converted into octahedral species by additional coordination of tri-methyl phosphate molecules. A monobromo-complex has only been found for cobalt, namely $[CoBr(TMP)_5]^+$, and is hexacoordinated.

Table 75. *Physical Properties of Trimethyl Phosphate*

Melting Point (°C)	−46.1
Boiling Point (°C)	194.0
Density at 20° (g · cm⁻³)	1.2145
Viscosity at 20° (Centipoise)	2.32
Dielectric Constant at 20°	20.6
Specific Conductivity at 20° (Ohm⁻¹ · cm⁻¹)	$1.2 \cdot 10^{-9}$
Donor Number DN$_{SbCl_5}$	23.0

In the respective chloro-systems tetrahedral $MCl_2(TMP)_2$, $[MCl_3(TMP)]^-$ and $[MCl_4]^{--}$ units were detected for cobalt(II) and nickel(II) and it appears that autocomplex formation occurs to a certain degree, when the dichlorides are dis-solved in trimethyl phosphate[68]. It may be noted that $[CoCl(TMP)_5]^+$ has not been detected in TMP, while both $[CoBr(TMP)_5]^+$ and $[CoN_3(TMP)_5]^+$ are readily formed. Higher azide-coordinated species of cobalt(II) are tetrahedral. $[Co(N_3)_4]^{--}$ is practically undissociated in TMP while in acetonitrile dissociation takes place[70]. In the nickel(II) azidosystem octahedral hexaazidonickelate $[Ni(N_3)_6]^{4-}$ appears to be present according to spectral evidence but potentiometric and conducto-metric results indicate the formation of $Ni(N_3)_2$ and $[Ni(N_3)_4]^{2-}$ which also seem to be hexacoordinated[70].

In the copper(II) system chloride and azide ligands lead to the same coordina-tion types[53, 68] namely $[CuX]^+$, $[CuX_2]$, $[CuX_3]^-$ and $[CuX_4]^{--}$, which seem to have distorted tetrahedral structures. Both in the manganese(II)-chloro and in the manganese(II) azido systems hexacoordinated species have been found[78,79].

The vanadyl ion gives with chloride and azide ions both neutral compounds and anionic species[80], the latter being presumably $[VOCl_4]^{2-}$ and $[VO(N_3)_5]^{3-}$ respectively. Autocomplex formation seems to occur in these systems[80].

In the vanadium(III) systems no anionic complexes are found. While this is not surprising for a competitive ligand of low donor properties such as the bromide ion, the insolubilities of the trichloride and triazide seem to prevent the forma-tion of chloro- and azido-complex anions[81,82]. It may be assumed that the insoluble "trihalides" are polymeric and thus hexacoordinated. Similar considera-tions may be responsible for the absence of azidochromates—only the species $[Cr(N_3)(TMP)_5]^{2+}$ and $[Cr(N_3)_3]_x$ being present[82]—while a tetrachlorochro-mate(III) appears to exist in solutions of trimethyl phosphate containing chloride ions[71].

In the ferric chloride systems the tetrahedral species $[FeCl_2(TMP)_2]^+$, $FeCl_3(TMP)$ and $[FeCl_4]^-$ are present and it appears that $FeCl_3$ dissolved in trimethyl phosphate undergoes autocomplex formation[78,127] as is known to occur in triethyl phosphate[128]. With azide ions as competitive ligands octahedral units are produced[79].

Table 76. *Coordination Forms Found in Solutions of Trimethyl Phosphate*

System \ Competitive Anions coordinated	1	2	3	4	5	6	References
$Co^{++} + I^-$	−	+	−	+			71
$Co^{++} + Br^-$	+	+	+	+			66
$Co^{++} + Cl^-$	−	+	+	+			53, 68
$Co^{++} + N_3^-$	−	+	−	+			70
$Co^{++} + NCS^-$	+	−	+	+			71
$Ni^{++} + I^-$	−	−	−	−			72
$Ni^{++} + Br^-$	−	+	−	+			66
$Ni^{++} + Cl^-$	−	+	+	+			53, 68
$Ni^{++} + N_3^-$	−	+	−	+	−	+	70
$Ni + NCS^-$	−	+	−	+	−	+	72
$Cu^{++} + Cl^-$	+	+	+	+			53, 68
$Cu^{++} + N_3^-$	+	+	+	+			70
$Mn^{++} + Br^-$	−	−	−	−			77
$Mn^{++} + Cl^-$	−	+	−	+	−	+	78
$Mn^{++} + N_3^-$	−	+	−	−	−	+	79
$VO^{++} + Br^-$	−	+	−	+			80 a
$VO^{++} + Cl^-$	−	+	−	+	−		80
$VO^{++} + N_3^-$	−	+	−	+			80
$VO^{++} + NCS^-$	−	+	+	+			80 a
$Ti^{3+} + Cl^-$	−	+	+	?	−	+	81
$Ti^{3+} + N_3^-$	−	+	+	−	−	+	82
$V^{3+} + Br^-$	+	−	+	−	−		77
$V^{3+} + Cl^-$	+	−	+	−	−	−	81
$V^{3+} + N_3^-$	+	−	+	−	−	−	82
$Cr^{3+} + Cl^-$	+	−	+	+	−	−	81
$Cr^{3+} + N_3^-$	+	−	+	−	−	−	82
$Fe^{3+} + Cl^-$	−	+	+	+	−	−	78, 127
$Fe^{3+} + N_3^-$	−	−	+	−	−	+	79

The relative chloride ion affinities were investigated between $CoCl_2$, $NiCl_2$ and $CuCl_2$ on the one side and $SbCl_5$, $SnCl_4$ and $TiCl_4$ on the other side[53]. Whilst in acetonitrile in the $CoCl_2$-$TiCl_4$ system, Co^{++} is solvent-stabilized and $[TiCl_6]^{--}$ is produced, trimethyl phosphate stabilizes $[TiCl_3]^+$, since $TiCl_4$ acts as a chloride ion donor towards $CoCl_2$ in this solvent:

$$CoCl_2 + TiCl_4 \rightleftharpoons [CoCl_3]^- + [TiCl_3]^+$$

[127] GUTMANN, V., and G. HAMPEL: Mh. Chem. **94**, 830 (1963).
[128] MEEK, D. W., D. K. STRAUB, and R. S. DRAGO: J. Amer. Chem. Soc. **82**, 6013 (1960).

In the other systems investigated the reactions occurred in accordance with the equations[53]:

$$MCl_2 + SbCl_5 \rightleftharpoons [MCl]^+ + [SbCl_6]^- \quad \text{and}$$
$$MCl_2 + 2\,SbCl_5 \rightleftharpoons M^{++} + 2\,[SbCl_6]^-$$

The order of chloride affinity in trimethylphosphate was found[53]:

$$SbCl_5 \sim SnCl_4 > CoCl_2 > TiCl_4$$

The number of coordination forms found in trimethyl phosphate within a particular system is smaller than in acetonitrile or propanediol-1,2-carbonate[3]. Some of the lower X^--coordinated species show high kinetic stabilities. The tendency to form solvate bonds is reflected in autocomplex formation for many systems and in the ionization of manganese(II) bromide and nickel(II) iodide.

11. Tributyl Phosphate (TBP) ($DN_{SbCl_5} = 23.7$)

Tributyl phosphate is well known as an excellent extracting agent for heavy metal ions, since it dissolves many ionic compounds. Because of the high donor number of the solvent many ionic compounds are ionized, but electrolytic dissociations hardly take place due to the low dielectric constant. Tetraalkylammonium perchlorate gives nearly non-conducting solutions due to extensive association of the ions and polarographic investigations are impossible on such solutions[129].

Table 77. *Some Physical Properties of Tributyl Phosphate*

Melting Point (°C)	−80.0
Boiling Point (°C)	289.0
Density at 25° (g · cm⁻³)	0.9727
Viscosity at 20° (Centipoise)	3.89
Dielectric Constant at 20°	6.8
Specific Conductivity at 25° (Ohm⁻¹ · cm⁻¹)	$1.25 \cdot 10^{-6}$
Donor Number (DN_{SbCl_5})	23.7

The donor properties of the solvent molecules are high and similar to those of trimethyl phosphate but there are remarkable differences in the abilities of the two solvents to allow formation of halide complexes[123]. No complex formation was found to occur in tributyl phosphate between $(C_2H_5)_4NCl$ or $(C_6H_5)_3CCl$ and $FeCl_3$, $ZnCl_2$, BCl_3, $AlCl_3$ (very low solubility), $TiCl_4$, $SnCl_4$ or $SbCl_5$[130]. The structure of the zinc chloride compound was suggested as follows[131]:

[129] GUTMANN, V., and M. MICHLMAYR: Unpublished.
[130] BAAZ, M., V. GUTMANN, and J. R. MASAGUER: Mh. Chem. **92**, 590 (1961).
[131] SCHAARSCHMIDT, K., G. MENDE, and G. EMRICH: Z. Chem. (Leipzig) **6**, 275 (1966).

The extraction of cupric chloride from aqueous hydrochloric acid by tributyl phosphate has been studied by tracer techniques[132] and it was concluded that the species $[CuCl_3(TBP)_2]^-$ and $[CuCl_4]^{--}$ were involved.

Ferric chloride which is known to undergo autocomplex formation in trimethyl phosphate, but although being extensively ionized in TBP, the halide remains undissociated owing to the low dielectric constant of this solvent:

$$FeCl_3 + 6\,TBP \rightleftharpoons [Fe(TBP)_6]^{3+} + 3\,Cl^-$$

The solvated ferric ion is the only iron-containing species which is stable in the presence of chloride ions in the anhydrous solvent, although the formation of $[FeCl_4 \cdot TBP]^-$ has been claimed[133] in the presence of water.

This behaviour may be mainly due to both the steric properties of the solvent molecules (through complete shielding of the metal cations) and the low dielectric constant of the solvent, which will not allow the competitive ligands to be available in significant concentrations.

12. Dimethylformamide (DMF) $(DN_{SbCl_5} \sim 27)$

Dimethylformamide is an excellent solvent for various classes of compounds, which in solution are considerably dissociated due to the high dielectric constant. It is a good hydrogen bonding agent particularly for organic compounds which contain active hydrogen atoms.

Table 78. *Some Physical Properties of* DMF

Melting Point (°C)	−61.0
Boiling Point (°C)	153.0
Density at 25° $(g \cdot cm^{-3})$	0.9443
Viscosity at 25° (Centipoise)	0.796
Dielectric Constant at 25°	36.71
Specific Conductivity at 25° $(Ohm^{-1} \cdot cm^{-1})$	$2 \cdot 10^{-7}$
Donor Number DN_{SbCl_5}	27.0

Numerous adducts have been prepared with acceptor molecules[134] such as with SO_3, $SnCl_4$, $SnBr_4$, $AlCl_3$, $CdCl_2$ and $SbCl_5$ and these all give conducting solutions. $AlCl_3$ gives an extremely hygroscopic compound of composition $AlCl_3(DMF)_6$, while $SnCl_4$ is coordinated by two and $SbCl_5$ by one DMF-molecule in the solid state. It has been established that the O-atom acts as the donor atom in most adducts[135] and that protons are also coordinated to the oxygen atom of the solvent molecule[136].

[132] MORRIS, D. F. C., and E. R. GARDNER: Electrochim. Acta **8**, 823 (1963).

[133] GUENZLER, C., and P. MÜHL: Z. Chem. (Leipzig) **5**, 148 (1965).

[134] PAUL, R. C., S. SINGH, and B. R. SREENATHAN: Ind. J. Chem. **2**, 97 (1964); C. A. **61**, 1351 g (1964).

[135] GORE, E. S., D. J. BLEARS, and S. S. DANYLUK: Can. J. Chem. **43**, 2135 (1965).

[136] KUHN, S. J., and J. S. McINTYRE: Can. J. Chem. **43**, 997 (1965).

Only very strong donor solvents are capable of dissolving compounds consisting of metal clusters, such as molybdenum dichloride Mo_6Cl_{12}, which is insoluble in most other solvents. Dimethylformamide and the somewhat stronger donor solvent dimethyl sulphoxide give the adducts[137] $Mo_6Cl_{12}(DMF)_2$ and $Mo_6Cl_{12}(DMSO)_2$ respectively. With silver perchlorate an ionic compound is formed which leaves the metal cluster $[Mo_6Cl_8]^{4+}$ intact:

$$Mo_6Cl_{12}(DMF)_2 + 4\,AgClO_4 + 4\,DMF \rightleftharpoons [Mo_6Cl_8(DMF)_6][ClO_4]_4 + 4\,AgCl$$

Even polymeric molybdenum(III) bromide is soluble through complex formation[138,139].

$$(MoBr_3)_n + 3\,n\,DMF \rightleftharpoons n\,MoBr_3(DMF)_3$$

Cobalt(II) chloride undergoes autocomplex formation in DMF[13]:

$$2\,CoCl_2 + 6\,DMF \rightleftharpoons [Co(DMF)_6]^{2+} + [CoCl_4]^{--}$$
$$[CoCl_4]^{--} + DMF \rightleftharpoons [CoCl_3DMF]^- + Cl^-$$
$$[Co(DMF)_6]^{2+} + Cl^- \rightleftharpoons [Co(DMF)_5Cl]^+ + DMF$$

The compound $CoCl_2(DMF)_2$ was isolated and was found to be different from tetrahedral species $CoCl_2D_2$ according to the reflectance spectra; it was tentatively formulated as $[Co(DMF)_6][Co_2Cl_6]$ containing a dimeric cobalt chloride unit.

Table 79. *Coordination Forms of* Co^{++}, Ni^{++} *and* Cu^{++} *in DMF*

System	Cl- coordinated	1	2	3	4	References
$Co^{++} + Cl^-$		+	+	+	+	13
$Ni^{++} + Cl^-$		+	−	+	?	140
$Cu^{++} + Cl^-$		+	−	+	−	140

Autocomplex formation is even more evident in the nickel(II) and copper(II) chloro-systems, where only $[MCl]^+$ and $[MCl_3]^-$ were detected[140]:

$$2\,NiCl_2 + 4\,DMF \rightleftharpoons [NiCl(DMF)_3]^+ + [NiCl_3(DMF)]^-$$
$$2\,CuCl_2 + 4\,DMF \rightleftharpoons [CuCl(DMF)_3]^+ + [CuCl_3(DMF)]^-$$

Proton n.m.r. studies of solutions of anhydrous perchlorate salts of transition metal ions indicate that the primary solvation spheres of the cations Ni^{2+}, Co^{2+}, Cu^{2+}, Fe^{3+} and Cr^{3+} are kinetically well defined[141,142]. Proton n.m.r. spectra of solutions of $Al(DMF)_6(ClO_4)_3$ in DMF indicate the non-equivalence of DMF molecules in the coordination sphere[142].

[137] COTTON, A. F., and N. F. CURTIS: Inorg. Chem. **4**, 241 (1965).

[138] KOMORITA, T., S. MIK, and S. YAMADA: Bull. Chem. Soc. Japan **38**, 123 (1965); C. A. **62**, 11401 f (1965).

[139] ALLEN, E. A., K. FEENAN, and G. W. A. FOWLES: J. Chem. Soc. **1965**, 1636.

[140] HUBACEK, H., B. STANčIć, and V. GUTMANN: Mh. Chem. **94**, 1118 (1963).

[141] MATWIYOFF, N. A.: Inorg. Chem. **5**, 788 (1966).

[142] MOVIUS, W. G., and N. A. MATWIYOFF: Inorg. Chem. **6**, 847 (1967).

Dimethylformamide has been used as a solvent for iodides[143]. Class (b) metal iodides, such as mercuric iodide, are undissociated in DMF; mercuric iodide is dimeric in dimethylformamide solution with the metal-iodine bonds stronger than metal-DMF bonds. On the other hand iodides of class (a) metals, such as lead iodide, bismuth iodide or stannic iodide, are considerably dissociated and thus provide iodide ions[143] which can be coordinated to mercuric iodide with formation of strong metal-iodine bonds:

$$BiI_3 + 3\,HgI_2 \rightleftharpoons Bi[HgI_3]_3$$

13. N,N-Dimethylacetamide (DMA) (DN$_{SbCl_5}$ = 27.8)

N,N-Dimethylacetamide is an excellent ionizing solvent with high donor properties. Many salts will undergo autocomplex formation and some of them, such as iodides and bromides of class (a) transition elements are easily ionized without showing tendencies to form anionic complexes.

Table 80. *Physical Properties of* DMA

Melting Point (°C)	+20.0
Boiling Point (°C)	+165.0
Density at 25° (g · cm^{-3})	0.9366
Viscosity at 25° (Centipoise)	0.919
Dielectric Constant at 25°	37.8
Specific Conductivity at 25° (Ohm^{-1} · cm^{-1})	10^{-7}
Dipole Moment (Debye)	3.79 (in benzene)
Donor Number DN$_{SbCl_5}$	27.8

Since N,N-dimethylacetamide has more favourable steric properties than tributyl phosphate, ferric chloride is converted into tetrachloroferrate(III) in the presence of chloride ions in N,N-dimethylacetamide; ferric chloride also undergoes considerable autocomplex formation in the pure solvent[144,145]

$$2\,FeCl_3 + 4\,DMA \rightleftharpoons [FeCl_2(DMA)_4]^+ + [FeCl_4]^-$$

In the cobalt(II) chloro-system the neutral species is present only in very small amounts due to extensive autocomplex formation[13]:

$$2\,CoCl_2 + 6\,DMA \rightleftharpoons [CoCl(DMA)_5]^+ + [CoCl_3(DMA)]^-$$
$$DMA + [CoCl(DMA)_5]^+ \rightleftharpoons [Co(DMA)_6]^{++} + Cl^-$$
$$[CoCl_3(DMA)]^- + Cl^- \rightleftharpoons [CoCl_4]^{--} + DMA$$

while $[CoBr_4]^{--}$ and $[CoI_4]^{--}$ are unstable in this solvent. The highest halogen-coordinated species are $[CoCl_4]^{--}$, $[CoBr_3]^-$ and CoI_2, the latter being appreciably ionized in this solvent.

It will be of interest to extend investigations of other systems to this solvent.

---

[143] GAIZER, F., and M. T. BECK: J. Inorg. Nucl. Chem. **29**, 21 (1927).

[144] DRAGO, R. S., R. L. CARLSON, and D. HART: ref. in "Non-Aqueous Solvent Systems," Ed. T. C. WADDINGTON, Academic Press London-New York 1965.

[145] DRAGO, R. S., R. L. CARLSON, and K. F. PURCELL: Inorg. Chem. **4**, 15 (1965).

14. Dimethyl Sulphoxide (DMSO) $(DN_{SbCl_5} = 29.8)$

Dimethyl sulphoxide is a strong donor solvent with favourable sterical properties. It reacts therefore with practically all acceptors with considerable evolution of heat and numerous solvates are known[146,147]. Some of the adducts such as $SnCl_4(DMSO)_2$ and $FeCl_3(DMSO)_2$ show great thermal stabilities. Coordination towards class (a) metal ions occurs through the oxygen atom and in compounds of class (b) metal ions, such as $PdCl_2(DMSO)_2$, infrared evidence suggests $Pd \leftarrow S = O$ bonding[147]. Thus dimethyl sulphoxide may be considered as a solvent molecule with a "hard" (oxygen) and with a "soft" (sulphur) end.

Most of the transition metal cations are strongly bonded to six dimethyl sulphoxide molecules making ligand exchange reactions more difficult to accomplish than for example in acetonitrile. Iodides, bromides and many chlorides of class (a) metals such as $FeCl_3$[127,148] and $VOCl_2$[80] are considerably ionized in this solvent and the formation of the corresponding halide complex anions is not possible.

Fluoro-, azido-, cyano- and thiocyanato-complexes of class (a) metal ions are, however, accessible in this solvent and it may be expected that neutral compounds with strong donor molecules, such as pyridine, may be formed.

Since dimethyl sulphoxide has also a reasonably high dielectric constant, the ionized compounds are considerably dissociated. Due to its high donor properties dimethyl sulphoxide dissolves various classes of chemical compounds. Many organic compounds including polymers are soluble. Numerous reactions, such as elimination, dehydration, isomerization, substitution, condensation, addition and redox-reactions may be carried out in dimethyl sulphoxide solutions with great ease[149].

Table 81. *Physical Properties of Dimethyl Sulphoxide*

Melting Point (°C)	18.2
Boiling Point (°C)	189.0
Density at 20° $(g \cdot cm^{-3})$	1.100
Viscosity at 20° (Centipoise)	2.473
Dielectric Constant at 20°	48.9
Specific Conductivity at 25° $(Ohm^{-1} \cdot cm^{-1})$	$3 \cdot 10^{-8}$
Dipole Moment (Debye)	3.9
Donor Number (DN_{SbCl_5})	29.8
Enthalpy of Vaporization $(kcal \cdot mole^{-1})$	13.67
Trouton-Constant $(cal \cdot mole^{-1} \cdot deg^{-1})$	29.6

Due to the high donor properties of the solvent molecules halide ion-transfer reactions are limited in solutions of dimethyl sulphoxide. Although in the system $CoCl_2$—$TiCl_4$ chloride ion transfer gives in acetonitrile Co^{++} and $[TiCl_6]^{--}$ and in trimethylphosphate $[CoCl_3]^-$ and $[TiCl_3]^+$, no halide transfer is observed in dimethyl sulphoxide[53]. Complex iodides and complex bromides of class (a)

[146] SCHLÄFER, H. L., and W. SCHAFFERNICHT: Angew. Chem. **72**, 618 (1960).

[147] COTTON, F. A., and R. FRANCIS: J. Amer. Chem. Soc. **82**, 2986 (1960).

[148] GUTMANN, V., and L. HÜBNER: Mh. Chem. **92**, 126 (1960).

[149] MARTIN, D., A. WEISE, and H. J. NICLAS: Angew. Chem. **79**, 340 (1967); Int. Ed. **6**, 318 (1967).

metal ions are completely dissociated and thus have very low stabilities in liquid dimethyl sulphoxide. The adduct of cobalt(II) iodide has been formulated as $[Co(DMSO)_6]I_2$[147]. It can be seen from Table 82 that no bromide coordination has been found to occur to cobalt(II):

$$CoBr_2 + 6\,DMSO \rightleftharpoons [Co(DMSO)_6]^{++} + 2\,Br^-$$

but the compound with cobalt(II) bromide $CoBr_2(DMSO)_3$ has been formulated as $[Co(DMSO)_6][CoBr_4]$ in the solid state[147].

Table 82. *Coordination Forms in Solutions of Dimethyl Sulphoxide*

System	Competing Anions coordinated						References
	1	2	3	4	5	6	
$Co^{++} + I^-$	−	−	−	−			71
$Co^{++} + Br^-$	−	−	−	−			66
$Co^{++} + Cl^-$	−	+	−	+			53, 68
$Co^{++} + N_3^-$	−	+	−	+			70
$Co^{++} + NCS^-$	−	−	−	+			71
$Co^{++} + CN^-$	−	+	−	−	+		71
$Ni^{++} + I^-$	−	−	−	−			72
$Ni^{++} + Br^-$	−	−	−	−			66
$Ni^{++} + Cl^-$	−	−	−	−			53, 68
$Ni^{++} + N_3^-$	−	+	−	+	−		70
$Ni^{++} + NCS^-$	−	+	−	+	−	+	72
$Ni^{++} + CN^-$	−	+	−	+			72
$Cu^{++} + Cl^-$	+	+	−	−			53
$Cu^{++} + N_3^-$	+	+	−	+			70
$Mn^{++} + Br^-$	−	−	−	−			77
$VO^{++} + Br^-$	−	−	−	−			80 a
$VO^{++} + Cl^-$	+	−	−	−	−		80
$VO^{++} + N_3^-$	−	+	−	+			80
$VO^{++} + NCS^-$	+	−	−	+			80 a
$Cr^{3+} + Cl^-$	−	+	+	−			150
$V^{3+} + Br^-$	−	−	−	−			77
$Fe^{3+} + Cl^-$	−	−	−	−	−	−	127, 148, 151, 152
$Fe^{3+} + N_3^-$	−	−	−	+	−	−	151
$Fe^{3+} + F^-$	−	+	−	+	−	−	151
$Fe^{3+} + CN^-$	−	+	+	+	−	−	151
$Fe^{3+} + SCN^-$	−	−	−	−	−	+	151

In the chloro-system both tetrahedral $CoCl_2$ and $[CoCl_4]^{--}$ have been found and the occurrence of autocomplex formation is indicated. These findings in solution are also in accordance with the formulation of the solid solvate as $[Co(DMSO)_6][CoCl_4]$. The analogous coordination forms are found also in the azide systems. The

150 SCHLÄFER, H. L., and H. P. OPITZ: Z. Chem. (Leipzig) **2**, 216 (1962).

151 CSISZAR, B., V. GUTMANN, and E. WYCHERA: Mh. Chem. **98**, 12 (1967).

152 DRAGO, R. S., D. M. HART, and R. L. CARLSON: J. Amer. Chem. Soc. **87**, 1900 (1965).

azide compounds $Co(N_3)_2$ and $[Co(N_3)_4]^{--}$ are formed[70], which are more stable than the corresponding chloro-compounds. No indications have been found for the formation of $[CoN_3]^+$, which is known in water and in trimethyl phosphate[70]. Auto-complex formation seems to be involved in this system with only moderate stability of the tetraazidocomplex-ion:

$$2\,Co(N_3)_2 + 6\,DMSO \rightleftharpoons [Co(DMSO)_6]^{2+} + [Co(N_3)_4]^{2-}$$
$$[Co(N_3)_4]^{2-} + 6\,DMSO \rightleftharpoons [Co(DMSO)_6]^{2+} + 4\,N_3^-$$

No bromo-compounds are formed from nickel(II) perchlorate and excess bromide ions; nickel(II) bromide is completely ionized in dimethyl sulphoxide[72].

$$NiBr_2 + 6\,DMSO \rightleftharpoons [Ni(DMSO)_6]^{2+} + 2\,Br^-$$

Even nickel(II) chloride undergoes ionization[53, 68] but nickel(II) azide undergoes autocomplex formation[70]. Although four azide-units are coordinated at nickel(II) in the anionic azido complex the spectra suggest a hexacoordinated species apparently arising from additional solvent coordination at the apices of the octahedron.

$$2\,Ni(N_3)_2 + 8\,DMSO \rightleftharpoons [Ni(DMSO)_6]^{2+} + [Ni(N_3)_4(DMSO)_2]^{2-}$$

The low stabilities of the complex compounds of class (a) metal ions with weak ligands are also reflected in the shapes of the potentiometric titration curves, which show no inflexions in contrast to the results in acetonitrile or trimethyl phosphate, where the stabilities are much higher.

The stabilities are somewhat higher in the copper(II) systems where the formation of the azido-complexes can also be followed by the potentiometric method[70]. Copper(II) chloride is partly ionized, since only $[Cu(DMSO)_6]^{++}$, $[Cu(DMSO)_3Cl]^+$ and $[Cu(DMSO)_2Cl_2]$ can be detected[53] and no chlorocuprates(II) are formed in dimethyl sulphoxide. Evidence is available for the formation of $[Cu(N_3)(DMSO)_3]^+$, $[Cu(N_3)_2(DMSO)_2]$ and $[Cu(N_3)_4]^{--}$ but the latter is partly dissociated.

Manganese(II) bromide, vanadium(III) bromide, the respective iodides, vanadyl chloride[80] and vanadyl bromide[153] are all completely ionized in DMSO[77]. No evidence is available even for the presence of unionized $VOCl_2$.

$$VOCl_2 + n\,DMSO \rightleftharpoons [VOCl(DMSO)_n]^+ + Cl^-$$
$$[VOCl(DMSO)_n]^+ + DMSO \rightleftharpoons [VO(DMSO)_{n+1}]^{++} + Cl^-$$

Only $[VOCl]^+$ is present in the solutions apart from the solvated vanadyl ion[80]. On the other hand both the diazide, and an azido-complex which is possibly the pentaazido complex[80] $[VO(N_3)_5]^{3-}$ are formed.

Chromium(III) chloride gives with DMSO $CrCl_3(DMSO)_5$, which contains the hexacoordinated dichlorochromium species[150], e.g. $[Cl_2Cr(DMSO)_4]Cl \cdot DMSO$. This compound is converted by heating into $CrCl_3(DMSO)_3$.

No chloro-coordinated iron(III) species have been identified[127,148,151,152], although DRAGO and PURCELL[10] conclude, from a small increase in absorbance on addition of chloride ions to ferric chloride solutions, that $[FeCl]^{++}$ should

[153] GUTMANN, V., and H. LAUSSEGGER: Mh. Chem. in press.

be stable in dimethyl sulphoxide. In any case ferric chloride is ionized extensively in dimethyl sulphoxide and the formation of an anionic chloro-complex cannot take place.

$$FeCl_3 + 6\,DMSO \rightleftharpoons [Fe(DMSO_6)]^{3+} + 3\,Cl^-$$

A pseudooctahedral $[Fe(N_3)_4(DMSO)_2]^-$ ion is produced with excess of azide ions[151], but the last two dimethyl sulphoxide positions cannot be replaced by azide ions. Thus autocomplex formation is predicted for ferric azide, although the latter has not been described yet. Likewise no ferric fluoride is formed when fluoride ions are added to the solution of iron(III) perchlorate. Again autocomplex formation takes place[151]:

$$4\,Fe(N_3)_3 + 12\,DMSO \rightleftharpoons [Fe(DMSO)_6]^{3+} + 3\,[Fe(N_3)_4(DMSO)_2]^-$$
$$2\,FeF_3 + 6\,DMSO \rightleftharpoons [FeF_2(DMSO)_4]^+ + [FeF_4(DMSO)_2]^-$$

1 : 1-Charge-transfer complexes of DMSO with iodine cyanide[154] were observed in carbon disulphide and with iodine[155] in carbon tetrachloride, as well as with iodine(I) chloride.

The strong donor properties of the solvent allow dissolution of compounds which are insoluble in most other solvents; anhydrous molybdenum(III) bromide, for example, gives an adduct[138,139].

Like ammonia, dimethyl sulphoxide gives an ionic complex with diborane[156], namely

$$\left[\begin{array}{c} H \\ \diagdown \\ H \diagup \end{array} B \begin{array}{c} \diagup DMSO \\ \diagdown DMSO \end{array} \right]^+ [BH_4]^-$$

The stepwise replacement of water molecules by dimethyl sulphoxide molecules in $[Cr(OH_2)_6][ClO_4]_3$ has been shown to take place with intermediates of general formula $[Cr(OH_2)_n(DMSO)_{6-n}]^{3+}$. The stepwise formation constants were found to be similar in magnitude[157].

When cis- and trans-$[Co(en)_2Cl_2]ClO_4$ are dissolved in dimethyl sulphoxide, they isomerize in the presence of chloride ions and undergo solvolysis[158] to give $[Co(en_2)(CH_3)_2SOCl]^{2+}$.

Niobium(V) chloride is solvolysed to an oxychloride species[159]:

$$3\,(CH_3)_2SO + NbCl_5 = NbOCl_3[(CH_3)_2SO]_2 + (CH_3)_2SCl_2$$

GAIZER and BECK[143] reported the ionization of class (a) metal iodides in DMSO. Dissociation of CdI_2 and PbI_2 and, to a higher degree, of BiI_3 is found to take place. Dissociation of SbI_3 and SnI_4 seems to be nearly complete. Thus the

[154] AUGDAHL, E., and P. KLAEBOE: Acta Chem. Scand. **18**, 18 (1964).

[155] KLAEBOE, P.: Acta Chem. Scand. **18**, 27 (1964).

[156] ACHRAN, G. E., and S. G. SHORE: Inorg. Chem. **4**, 125 (1965).

[157] ASHLEY, K. R., R. E. HAMM, and R. H. MAGNUSON: Inorg. Chem. **6**, 413 (1967).

[158] TOBE, M. L., and D. W. WATTS: J. Chem. Soc. **1964**, 2991.

[159] COPLEY, D. B., F. FAIRBROTHER, H. K. GRUNDY, and A. THOMPSON: J. Less Common Metals **6**, 407 (1964).

solutions contain free iodide ions which can be used to form strong iodo-complexes with iodides of class (b) metals[143], which are non-ionized in DMSO:

$$MI_n + n\,HgI_2 = M^{n+} + n\,[HgI_3]^-$$

At higher concentrations BEER's law is not valid and heteropolynuclear complexes have been assumed to be formed[143].

Certain gases, such as carbon dioxide, sulphur dioxide or the oxides of nitrogen are easily soluble in DMSO and may be determined by the polarographic method in this solvent[160–162].

Dimethyl sulphoxide is a useful reaction medium in organometallic synthesis. For example potassium hydroxide acts both as a protonating and dehydrating agent in dimethyl sulphoxide to enable the formation of ferrocene from hydrated ferrous chloride and cyclopentadiene[163].

$$8\,KOH + 2\,C_5H_6 + FeCl_2 \cdot 4\,H_2O \rightleftharpoons Fe(C_5H_5)_2 + 2\,KCl + 6\,KOH \cdot H_2O$$

Nickelocene and cobaltocene can be prepared in good yield in an analogous manner[163].

Numerous redox reactions have been reviewed elsewhere[164].

15. Hexamethylphosphoramide (HMPA) $(DN_{SbCl_5} = 38.8)$

The extremely high donor number suggests that HMPA will be a powerful ionizing solvent. On the other hand shape and size of the pyramidal solvent molecule

$$
\begin{array}{c}
(CH_3)_2N \\
(CH_3)_2N \!-\! P \rightarrow O \\
(CH_3)_2N
\end{array}
$$

will give rise to steric hindrance in highly solvated species and thus to complexes of unusual stereochemistry.

Table 83. *Some Physical Properties of* HMPA

Melting Point (°C)	7.2
Boiling Point (°C)	230–232/739 Torr
Density (g · cm⁻³)	1.02
Donor Number DN_{SbCl_5}	38.8
Dielectric Constant at 4°	33.5
Dipole Moment in Benzene at 25° (Debye)	5.54

It dissolves ionic compounds with great ease. Numerous metal ions, such as Co^{2+} and Ni^{2+}, which are known to give hexacoordinated species in most solvents, give tetracoordinated adducts in HMPA[165]. The tetrasolvates of the perchlorates

[160] DEHN, H., V. GUTMANN, H. KIRCH, and G. SCHÖBER: Mh. Chem. 93, 1348 (1962).
[161] GRITZNER, G., V. GUTMANN, and G. SCHÖBER: Mikrochim. Acta 1964, 193.
[162] GUTMANN, V.: Mikrochim. Acta 1966, 795.
[163] JOLLY, W. L.: private communication.
[164] EPSTEIN, W. W., and F. W. SWEAT: Chem. Revs. 67, 247 (1967).
[165] DONOGHUE, J. T., and R. S. DRAGO: Inorg. Chem. 1, 866 (1962).

of zinc, cobalt(II) and nickel(II) have been isolated from its solutions. The nickel complex is the most nearly tetrahedral cationic complex of nickel(II) known. The spectral and magnetic data for $[Fe(HMPA)_4]^{2+}$ also support a tetrahedral configuration for this cation and the X-ray powder pattern indicates isomorphism with known tetrahedral structures[166] of Zn^{2+}, Co^{2+} and Ni^{2+}. Fourcoordinate species have also been found for Mg^{2+}, Ca^{2+}, Ba^{2+} and Mn^{2+} ions. $[Al(HMPA)_4]^{3+}$ is nearly as tetrahedral as $[AlCl_4]^-$.

Infrared data indicate that the oxygen atom of HMPA is the donor atom in all these complexes. The position of HMPA in the spectrochemical series is believed to be between chloride and azide ions.

With dihalides of cobalt(II) and nickel(II), such as the chlorides, bromides and iodides, as well as with the thiocyanates, tetrahedral complexes with general formula $MeX_2 \cdot (HMPA)_2$ are formed[167]. As may be expected these compounds are non-conductors in a solvent of very low donor number, such as nitromethane. Likewise the disolvates of certain metal nitrates, such as cobalt(II), nickel(II) or copper(II) appear to have the nitrate groups located within the coordination spheres. The compound $Co(NO_3)_2 \cdot (HMPA)_2$ contains Co(II) in pseudotetrahedral C_{2v} environment. $Cu(NO_3)_2(HMPA)_2$ is thought to be a planar complex, while in $Ni(NO_3)_2(HMPA)_2$ a distorted tetrahedral structure appears to be present[167].

HMPA forms strong adducts with acceptor molecules, such as $SbCl_5$[168] and $(CH_3)_3SnCl$[169] and calorimetric measurements of the interactions with both of these acceptors show that HMPA is a much stronger donor solvent than DMSO. The contact shifts of the tetrahedral complexes of several transition metal ions were interpreted as indicating an increase in covalent bonding ability[170] in the series: $Mn(II) < Fe(II) < Co(II) < Ni(II) < Cu(II)$.

According to the high donor number of HMPA the formation of complex bromides and iodides appears to be unlikely in this solvent, but complexes with very strong competitive ligands may be accessible. Ionization of most halides, or at least considerable autocomplex formation is expected.

Another interesting feature in HMPA is its ability to dissolve alkali metals[171]; the blue solutions are strong reducing agents.

An excellent review[172] covers many aspects of the chemistry in HMPA, such as reactions with electrophiles, with nucleophiles, catalytic properties, isomerization, polymerization and substitution reactions.

[166] DONOGHUE, J. T., and R. S. DRAGO: Inorg. Chem. 2, 1158 (1963).

[167] DONOGHUE, J. T., and R. S. DRAGO: Inorg. Chem. 2, 572 (1963).

[168] GUTMANN, V., and A. SCHERHAUFER: Mh. Chem., 99, 335 (1968).

[169] BOLLES, T. F., and R. S. DRAGO: J. Am. Chem. Soc. 88, 3921 (1966).

[170] WAYLAND, B. B., and R. S. DRAGO: J. Amer. Chem. Soc. 87, 2372 (1965).

[171] CURIGNY, T., J. NORMANT, and H. NORMANT: Compt. Rend. Acad. Sci. Paris 258, 3503 (1964).

[172] NORMANT, H.: Angew. Chem. 79, 1029 (1967); Int. Ed. 6, 1046 (1967).

Chapter VIII

Coordination Chemistry of Certain Transition Metal Ions in Donor Solvents

Consecutive replacement of solvent molecules by competitive ligands in the coordination sphere of class (a) transition metal cations has been followed by spectrophotometric, potentiometric and conductometric techniques. Perchlorates or tetrafluoroborates have been often used, allowing the metal ions to achieve hexacoordination by solvent molecules in the non-aqueous solutions; competitive (anionic) ligands were added to the solutions in the form of tetraalkylammonium salts.

The equilibria involved in such competition reactions were described in the previous section and we may now attempt to discuss the influences of the solvents on the equilibria. Although kinetic aspects should not be ignored, we shall choose not to consider them at this stage.

1. Iodide Ions as Competitive Ligands

Metal-iodine bonds are much weaker for class (a) metals, than for class (b) metals. Thus class (a) metal-iodides are formed in solvents of low donor number whilst in solvents of high donor number, complete ionization will take place. Iodo-complexes of such metals are, therefore, unstable in solvents of high donor number.

Cobalt(II) iodide is subject to autocomplex formation in TMP and it is ionized in TMP, DMF, DMA and DMSO[1]. Nickel iodide is completely ionized in TMP and DMSO[2]. Tin(IV) iodide, bismuth(III) iodide, antimony(III) iodide, lead(II) iodide and cadmium(II) iodide are considerably ionized in DMF and completely ionized in DMSO [3].

On the other hand class (b) metal iodides are unionized even in solvents of high donor number because the metal-iodine bonds are considerably stronger than the metal-solvent bonds, and they are capable of accepting iodide ions in such solvents. It is therefore possible to produce complex iodides of class (b) elements in solvents of high donor number by providing iodide ions. GAIZER and BECK[3]

---

[1] GUTMANN, V., and O. BOHUNOVSKY: Mh. Chem. in press (1968). F. A. COTTON, and R. FRANCIS: J. Am. Chem. Soc. **82**, 2986 (1960).

[2] GUTMANN, V., and H. BARDY: Mh. Chem. in press (1968).

[3] GAIZER, P., and M. T. BECK: J. Inorg. Nucl. Chem. **29**, 21 (1967).

have produced several iodomercurates in this way. They have shown that mercuric iodide, which is unionized in DMSO and dimeric in DMF, will react with the iodides of lead, antimony or bismuth in these solvents to give compounds containing the HgI_3^- ion:

$$BiI_3 + 3\,HgI_2 \rightleftharpoons Bi(HgI_3)_3$$
$$SnI_4 + 4\,HgI_2 \rightleftharpoons Sn(HgI_3)_4$$

It may be expected that many other complex iodides can easily be obtained by this method. However, in concentrated solutions, which do not obey BEER's law, heteropolynuclear species appear to be formed[3].

On the other hand a solvent of low donor number such as nitromethane will allow the formation of $[CoI_4]^{--}$ ions even if only stoichiometric amounts of iodide ions are available. Such solvents may therefore be considered as levelling solvents with respect to the relative stabilities of analogous complex species containing different ligands.

2. Bromide Ions as Competitive Ligands

The bromide ion is a stronger ligand than the iodide ion towards class (a) metal ions, but metal-bromine bonds are still weak enough to allow complete ionization in strongly donating solvents; bromo-complexes are nevertheless easily accessible in solvents of low or medium donor number.

For cobalt(II), nickel(II), manganese(II) and vanadium(III) bromide analogous behaviour was found in the respective solvents:

In acetonitrile for cobalt(II)[4], nickel(II)[4] and manganese(II)[5] the following species were found: $[MBr]^+$, MBr_2, $[MBr_3]^-$ and $[MBr_4]^{2-}$. The monobromo-complex of cobalt seems to be octahedral, but all the other bromo-complexes appear to be tetrahedral or slightly distorted tetrahedral. For vanadium(III) $[VBr_2]^+$, VBr_3 and $[VBr_4]^-$ have been found in acetonitrile and it is reasonable to assume a certain amount of autocomplex formation in such solutions of the bromides.

In propanediol-1,2-carbonate no asymmetric units are found since only $[MBr_4]^{--}$ and $[VBr_4]^-$, respectively were detected. Thus autocomplex formation of the bromides is complete in this solvent. The interaction of propanediol-1,2-carbonate with transition metal cations appears stronger than might be expected from a consideration of its donor number, which is only slightly higher than that of acetonitrile; in this latter solvent many more coordination forms are produced.

$$2\,MBr_2 \rightleftharpoons M^{2+} + [MBr_4]^{2-}$$
in acetonitrile
$$2\,MBr_2 \rightleftharpoons M^{2+} + [MBr_4]^{2-}$$
complete in PDC

The high dielectric constant of PDC and sterical factors, are probably very helpful in autocomplex formation.

[4] GUTMANN, V., and K. FENKART: Mh. Chem. **98**, 1 (1967).
[5] GUTMANN, V., and K. FENKART: Mh. Chem. **98**, 286 (1967).

In acetone—a solvent of higher donor number, but with smaller and less bulky solvent molecules—$CoBr_2$, $[CoBr_3]^-$ and $[CoBr_4]^{--}$ were detected. Autocomplex formation is again indicated, and is particularly evident for nickel(II) bromide:

$$NiBr_2 \rightleftharpoons NiBr^+ + NiBr_3^-$$

The stability of the $[CoBr_4]^{--}$ is lower in acetone than in AN and PDC which have lower donor numbers.

The high donor number of trimethylphosphate promotes ionization of manganese(II) bromide, and no bromomanganate is formed with excess bromide ions:

$$MnBr_2 + 6\,TMP \rightleftharpoons [Mn(TMP)_6]^{2+} + 2\,Br^-$$

The same behaviour is expected in still stronger donor solvents.

Vanadium(III) bromide can exist in TMP, but it is partly ionized

$$VBr_3(TMP)_3 + 2\,TMP \rightleftharpoons [VBr(TMP)_5]^{++} + 2\,Br^-$$
$$[VBr(TMP)_5]^{++} + TMP \rightleftharpoons [V(TMP)_6]^{3+} + Br^-$$

and no bromo-complex can be formed.

For cobalt(II) and nickel(II) the tetrabromo-complexes can be formed, but they are highly dissociated:

$$[CoBr_4]^{2-} + 2\,TMP \rightleftharpoons CoBr_2(TMP)_2 + 2\,Br^-$$
$$CoBr_2(TMP)_2 + 4\,TMP \rightleftharpoons [Co(TMP)_6]^{2+} + 2\,Br^-$$

Complete ionization with no tendency to form any bromide coordinated species is found for all bromides in solutions in the strong donor-solvent dimethyl sulphoxide. Thus stabilities of tetrabromo-complexes of Ni(II), Co(II) and Cu(II) 6, 7 decrease with increasing donor number of the solvent:

$$AN > PDC > Acetone > TMP > DMSO$$

This series is further illustrated by manganese(II) bromide, which shows partial autocomplex formation in AN, complete autocomplex formation in PDC, and complete ionization into Mn^{++} and $2\,Br^-$ both in TMP and in DMSO.

We may thus briefly summarize the situation by reiterating that autocomplex formation increases according to the order $AN < PDC < (CH_3)_2CO < TMP$, bromometallate stabilities decrease in the order $AN > PDC > (CH_3)_2CO > TMP > DMSO$; and that ionization is more pronounced in DMSO than in TMP.

Redox reactions involving ferric and cupric ions were found in several solvents. Class (b) metal bromides, such as $HgBr_2$ are expected to remain essentially unionized in a solvent of high DN_{SbCl_5}, such as DMSO, and bromo-complexes of such metals may be accessible in this solvent.

3. Chloride Ions as Competitive Ligands

The chloride ion is a stronger ligand towards class (a) acceptors than the bromide ion. Thus chloro-complexes will be stable in certain solvents, in which bromo-complexes cannot be obtained due to the higher stabilities of solvent-coordinated species. For example cobalt(II) bromide is completely ionized in

[6] FINE, D. A.: J. Amer. Chem. Soc. **84**, 1139 (1962).

[7] SRAMKO, T.: Chem. Zvesti **17**, 725 (1963); C. A. **60**, 8697 b (1964).

DMSO, but tetrachlorocobaltate(II) is capable of existence in this solvent, in which $CoCl_2$ undergoes considerable autocomplex formation. Likewise manganese bromide is completely ionized in TMP, but $MnCl_2$ and $[MnCl_4]^{--}$ are known to exist in TMP solutions again with autocomplex formation of $MnCl_2$.

The solubilities of cobalt (II) and nickel(II) chlorides increase, in a qualitative way, with increasing donor number of the solvent[8]:

$$C_2H_4Cl_2 < NM < AN < PDC < TMP < DMSO$$

Cobalt(II) chloride is slightly soluble in nitromethane ($DN_{SbCl_5} = 2.7$) to give non-conducting solutions containing $CoCl_2(NM)_2$. This compound is not ionized in the solution because of the low donor properties of the solvent and hence dissociation cannot take place although the solvent has a reasonably high dielectric constant. Addition of chloride ions leads easily to chlorocobaltates.

Both in acetonitrile[8] ($DN_{SbCl_5} = 14.1$) and acetone[6] ($DN_{SbCl_5} = 17$) the tetrahedral forms $CoCl_2$, $CoCl_3$ and $[CoCl_4]^{--}$ are produced. In trimethyl phosphate ($DN_{SbCl_5} = 23$) autocomplex formation of cobalt(II) chloride is indicated[8]:

$$2\,CoCl_2 + 6\,TMP \rightleftharpoons [Co(TMP)_6]^{2+} + [CoCl_4]^{--}$$

and this appears to take place to a larger extent in DMF[9] ($DN_{SbCl_5} = 27$) and DMA[10] ($DN_{SbCl_5} = 27.8$). In the still stronger donor solvent DMSO ($DN_{SbCl_5} = 29.8$) this reaction is complete and $CoCl_2(DMSO)_2$ has not been detected in the solution[11]. The order of stability of $[CoCl_4]^{2-}$ in these solvents, namely

$$C_2H_4Cl_2 > NM > AN > PDC > TMP > DMF > DMA \gtrless DMSO$$

is in accord with the inverse order of donor numbers[12,13] of the respective solvent molecules[8]. On the other hand ionization is favoured by a solvent of high DN_{SbCl_5}. Cobalt(II) chloride is unionized in nitromethane and is known to undergo autocomplex formation in PDC and TMP, while in DMSO partial ionization takes place.

Nickel(II) chloride has somewhat lower solubilities than cobalt(II) chloride. $NiCl_3^-$ and $NiCl_4^{--}$ are produced in acetonitrile and also in trimethyl phosphate where autocomplex formation is appreciable[8]. In tributyl phosphate the units were not identified with certainty[14], but very low stabilities of the chloride-coordinated nickel species are indicated.

In DMF ($DN_{SbCl_5} = 27$) ionization of nickel(II) chloride is observed with some autocomplex formation[15], but tetrachloronickelate(II) is not stable and in dimethyl sulphoxide ($DN_{SbCl_5} = 29.8$) ionization of nickel(II) chloride is considerable.

$$2\,NiCl_2 + 6\,TMP \rightleftharpoons [Ni(TMP)_6]^{2+} + [NiCl_4]^{2-}$$
$$NiCl_2 + 5\,DMF \rightleftharpoons [Ni(DMF)_5Cl]^+ + Cl^-$$
$$[Ni(DMF)_5Cl]^+ + DMF \rightleftharpoons [Ni(DMF)_6]^{2+} + Cl^-$$
$$[Ni(DMF)_5Cl]^+ + 2\,Cl^- \rightleftharpoons [Ni(DMF)_3Cl_3]^- + 2\,DMF$$
$$NiCl_2 + 6\,DMSO \rightleftharpoons [Ni(DMSO)_6]^{2+} + 2\,Cl^-$$

[8] Baaz, M., V. Gutmann, G. Hampel and J. R. Masaguer: Mh. Chem. **93**, 1416 (1962).
[9] Buffagny, S., and T. M. J. Dunn: J. Chem. Soc. **1961**, 5105.
[10] Drago, R. S., and K. F. Purcell: Chapter 5 in "Non-Aqueous Solvent Systems," Ed. T. C. Waddington, Academic Press, London, New York 1965.
[11] Gutmann, V., and L. Hübner: Mh. Chem. **92**, 1261 (1961).
[12] Gutmann, V., and E. Wychera: Inorg. Nucl. Chem. Letters **2**, 257 (1966).
[13] Gutmann, V., and E. Wychera: Rev. Chim. Min. **3**, 941 (1966).
[14] Katzin, L. I.: Nature **182**, 1013 (1958).
[15] Hubacek, H., B. Stančić, and V. Gutmann: Mh. Chem. **94**, 1118 (1963).

For copper(II) chloride the behaviour is similar: every possible chloro-complex is found in AN and TMP with the probability of autocomplex formation of $CuCl_2$, which is considerable in DMF. In DMSO ionization takes place and no chloro-complexes are produced, whilst in TMP, $CuCl_3^-$ and $CuCl_4^{--}$ are formed; the latter is unstable in both DMF and DMSO and both $CuCl_3^-$ and $CuCl_4^{--}$ are unstable in DMSO.

In the manganese(II) bromide systems the solvate bonds appear to be stronger than in the corresponding bromo-systems of cobalt(II), nickel(II) and copper(II). The latter undergo autocomplex formation in TMP, whereas manganese(II) bromide is completely ionized in this solvent. The same trend is also found in the manganese(II) chloride systems. Manganese(II) chloride undergoes autocomplex formation in AN and to a larger extent in PDC and TMP. Its behaviour in DMSO has not been investigated.

Vanadium(III) gives in AN with chloride ions[16],[17] $VCl_3(AN)_3$ and $[VCl_4(AN)_2]^-$. Autocomplex formation seems to be involved in PDC and ionization of the trichloride in TMP, where no anionic chloro-complex has been detected[17]. The behaviour of titanium(III) is analogous[17].

Chromium(III) chloride gives in AN and PDC $[CrCl_2D_4]^+$ units as well as $CrCl_3D_3$ and $[CrCl_4D_2]^-$, but in TMP it undergoes autocomplex formation without considerable ionization[17].

Ferric chloride gives several chloride coordinated species in nitromethane[18] and phosphorus oxychloride[19],[20] but the species are polymeric at $c \sim 10^{-3}M$ in the absence of chloride ion donors. Tetrachloroferrate is formed by addition of chloride ions in these solvents as well as in AN and PDC[21]. Autocomplex formation is found to take place in propanediol-1,2-carbonate[21], in triethyl phosphate[22] and in TMP[21],[23] as well as in DMA[10]. In N-methylacetamide[10] (NMA) which has similar donor properties to the latter, in tributyl phosphate ($DN_{SbCl_5} \sim 24.5$) and in DMSO ($DN_{SbCl_5} = 29.8$) no chloride-coordinated species are observed even in the presence of excess chloride ions:

$$FeCl_3 + 6\,TBP \rightleftharpoons [Fe(TBP)_6]^{3+} + 3\,Cl^-$$
$$FeCl_3 + 6\,NMA \rightleftharpoons [Fe(NMA)_6]^{3+} + 3\,Cl^-$$
$$FeCl_3 + 6\,DMSO \rightleftharpoons [Fe(DMSO)_6]^{3+} + 3\,Cl^-$$
$$2\,FeCl_3 + 4\,DMA \rightleftharpoons [Fe(DMA)_4Cl_2]^+ + [FeCl_4]^-$$

A consideration of the behaviour of ferric chloride in these strong donor solvents shows that other factors come into play in addition to the role of the donor number. Tributyl phosphate has unfavourable steric properties and stabilizes the symmetrical $[Fe(TBP)_6]^{3+}$-ion [24]. N-methylacetamide does not favour the formation of the $FeCl_4^-$-ion, which is also unstable in water and in the alcohols. The order of stabilities of $FeCl_4^-$ ions in different solvents is approximately:

$$NM > AN > PDC > TMP > TEP > DMA > H_2O > NMA > TBP > DMSO$$

[16] GUTMANN, V., G. HAMPEL, and W. LUX: Mh. Chem. **96**, 593 (1965).
[17] GUTMANN, V., A. SCHERHAUFER, and H. CZUBA: Mh. Chem. **98**, 619 (1967).
[18] DE MAINE, P. A. D., and E. KOUBEK: J. Inorg. Nucl. Chem. **11**, 329 (1959).
[19] GUTMANN, V., and M. BAAZ: Mh. Chem. **90**, 729 (1959).
[20] GUTMANN, V., M. BAAZ, and L. HÜBNER: Mh. Chem. **91**, 537 (1960).
[21] GUTMANN, V., and W. K. LUX: Mh. Chem. **98**, 276 (1967).
[22] MEEK, D. W., and R. S. DRAGO: J. Amer. Chem. Soc. **83**, 4322 (1961).
[23] GUTMANN, V., and G. HAMPEL: Mh. Chem. **94**, 830 (1963).
[24] BAAZ, M., V. GUTMANN, and J. R. MASAGUER: Mh. Chem. **92**, 590 (1961).

The highest chloride-coordinated vanadyl species is $[VOCl_4]^{2-}$ in AN, PDC, H_2O and TMP[25]. Extensive autocomplex formation is found in TMP and again complete ionization in DMSO, in which anionic chloro-complexes of VO^{2+} are unstable[25]:

$$VOCl_2 + 4\,DMSO \rightleftharpoons [VO(DMSO)_4Cl]^+ + Cl^-$$
$$[VO(DMSO)_4Cl]^+ + DMSO \rightleftharpoons [VO(DMSO)_5]^{++} + Cl^-$$

4. Azide Ions as Competitive Ligands

The azide ion has favourable steric properties and is a stronger ligand than the chloride ion. It may therefore be expected that most of the azido-complexes will also be accessible in strong donor solvents, such as dimethyl sulphoxide, where autocomplex formation is likely to occur. Ionization of azides may occur in solvents of $DN_{SbCl_5} > 30$.

Pseudooctahedral $Mn(N_3)_2$ and $[Mn(N_3)_4]^{2-}$ units are found in AN, PDC and TMP[26]. The diazides of cobalt(II) are tetrahedral in AN, TMP and DMSO[27]. The cationic monoazido-complex has been found only in TMP, namely $[Co(N_3)(TMP)_5]^+$, while tetrahedral tetraazidocobaltates are formed in AN, TMP and DMSO, in which considerable autocomplex formation is apparent:

$$2\,Co(N_3)_2 + 6\,DMSO \rightleftharpoons [Co(DMSO)_6]^{2+} + [Co(N_3)_4]^{2-}$$

Hexacoordination is maintained in the nickel-azide system in acetonitrile and in trimethyl phosphate, where species of general formula $[Ni(N_3)_aD_{6-a}]^{+2-a}$ are present. The azido-complexes show lower stabilities in DMSO than in AN since reasonable competition is provided by the strong donor solvent molecules[27, 28]. $[Ni(N_3)_4]^{2-}$ seems to have a distorted octahedral arrangement with 2 DMSO molecules coordinated at the apices of the octahedron. Thus $[Ni(N_3)_4]^{2-}$ may be considered as planar with respect to azide-coordination, as is known in $[Ni(CN)_4]^{2-}$. The preferred hexacoordinated arrangements in all three solvents may be favoured by the small size of the azide ions and by π-bonding contributions in the metal azide bonds[27].

In the copper(II) azido-system the distorted tetrahedral species $[CuN_3]^+$, $[Cu(N_3)_2]$, $[Cu(N_3)_3]^-$ and $[Cu(N_3)_4]^{2-}$ are found in AN and TMP. In DMSO the stabilities of the anionic complexes are lower and $[Cu(N_3)_3]^-$ has been found to be unstable in this solvent[27]. Autocomplex formation of copper(II) azide seems to occur in anhydrous dimethyl sulphoxide.

In the azido-systems of titanium(II), vanadium(III) and chromium(III) hexa-coordinated species are found[29]. Hexaazidovanadate(III) seems to be formed in AN and PDC, but in TMP the triazides are the highest azido-coordinated species,

[25] GUTMANN, V., and H. LAUSSEGGER: Mh. Chem. **98**, 439 (1967).

[26] GUTMANN, V., and W. K. LUX: J. Inorg. Nucl. Chem. **29**, 2391 (1967).

[27] GUTMANN, V., and O. LEITMANN: Mh. Chem. **97**, 926 (1966).

[28] GUTMANN, V.: Coord. Chem. Revs. **2**, 239 (1967).

[29] GUTMANN, V., O. LEITMANN, H. CZUBA, and A. SCHERHAUFER: Mh. Chem. **98**, 188 (1967).

which have been detected. These have low solubilities in this solvent possibly owing to polymeric structures in which hexacoordination by azide bridging may be present. The same features are observed in the chromium(III) azide system in TMP, while hexaazidochromate appears to be formed in AN and PDC. Again the stabilities of the anionic complexes are decreased with increasing donor properties of the solvents.

Iron(III) azide and an anionic azidoferrate(III) are formed in AN, PDC and TMP [26]. In PDC also $[FeN_3]^{++}$ and $[Fe(N_3)_2]^+$ units have been detected. In DMSO the only azide-coordinated species is $[Fe(N_3)_4(DMSO)_2]^-$. Thus ferric azide will undergo autocomplex formation and two solvent positions in the distorted octahedron cannot be replaced by the azide ions[30].

5. Thiocyanate Ions as Competitive Ligands

The thiocyanate ion is a stronger ligand than those discussed above. The influence of the donor number of the solvent on the stabilities of the complex ions will therefore be small. Indeed, in all solvents under consideration, the highest thiocyanate-coordinated forms are formed. Nickel(II)[31] gives in most solvents $Ni(NCS)_2$, $Ni(NCS)_4]^{2-}$ and also $[Ni(NCS)_6]^{4-}$, whilst with cobalt(II), tetrahedral $[Co(NCS)_4]^{2-}$ is found to exist as a stable anionic complex[32]. The small influence of the donor number of the solvent may be seen from the molar ratios $x = NCS^- : M^{2+}$ required to obtain the anionic complexes quantitatively.

Table 84. *Molar Ratios* $x = NCS^-: M^{++}$ *Required to Provide Quantitative Conversion into Anionic Complexes* 31, 32 $M^{++} = Ni^{++}$ *or* Co^{++} *respectively*

Solvent	DN_{SbCl_5}	x for $[Ni(NCS)_6]^{4-}$	x for $[Co(NCS)_4]^{2-}$
DMSO	29.8	500	200
DMA	27.8	250	13
TMP	23.0	250	5
PDC	15.1	70	7
AN	14.1	250	4
NM	2.7	?	4

6. Cyanide Ions as Competitive Ligands

Co^{2+} gives in NM[33], PDC[32] and DMSO[32] both the scarcely soluble dicyanide, and a pentacyano-complex $[Co(CN)_5]^{3-}$ which appears to have a pseudooctahedral structure by coordination of one solvent molecule. Ni^{2+} gives again scarcely soluble dicyanides and square planar tetracyanonickelates[31].

[30] CZISZAR, B., V. GUTMANN, and E. WYCHERA: Mh. Chem. **98**, 12 (1967).
[31] GUTMANN, V., and H. BARDY: Z. anorg. allg. Chem. in press (1968).
[32] GUTMANN, V., and O. BOHUNOVSKY: Mh. Chem. in press (1968).
[33] GUTMANN, V., and K. H. WEGLEITNER: Mh. Chem. **99**, 368 (1968).

7. An Attempt to Assign a Donor Number to Anions

The approach of measuring the enthalpy of the reaction in dichloroethane

$$SbCl_5 + D \rightleftharpoons D \cdot SbCl_5$$

cannot be applied to halide and pseudohalide anions[34], since ligand exchange reactions may take place and the solubilities of anion-donors are small in dichloroethane.

Equilibrium studies have been carried out in solutions of dichloromethane and acetonitrile, using vanadylacetylacetonate $VO(acac)_2$ as an acceptor[34] which has one ligand site available for coordination[35–37]. The V-O-bond energies of the chelate rings are comparable to the C-C-bond energy[38] and are stable enough to exclude ligand exchange reactions. The equilibria

$$VO(acac)_2 + X^- \rightleftharpoons [VOX(acac)_2]^-$$

were measured for $X^- = Cl^-$, NCS^- and N_3^- in dichloromethane and acetonitrile by the spectrophotometric method. The donor strength of the bromide ion was estimated, but data for fluoride- and cyanide-ions are not available yet. The neutral donor molecules TMP, DMF, DMSO, Py, $(C_6H_5)_3PO$ and HMPA have also been used. The following order was found both in dichloromethane and acetonitrile[34]: $Br^- < TMP < DMF \leq Cl^- < DMSO < (C_6H_5)_3PO < HMPA \sim Py < NCS^- < N_3^-$.

The series is in agreement with that of the donor numbers of the solvent molecules and also with the behaviour of competitive anions as described above[34].

8. Conclusion

It may be concluded that a solvent of low donor number acts as a levelling solvent for the stabilities of analogous complex species containing competitive ligands of different donor properties. Nitromethane allows the formation of anionic complexes of transition metal ions with I^-, Br^-, Cl^-, N_3^-, NCS^- nearly with stoichiometric amounts of the latter. On the other hand a solvent of high donor number may be considered to be a differentiating solvent with respect to the stabilities of analogous complex species with competitive ligands of different donor properties. Thus in DMSO complex iodides and bromides are unstable, while complex azides and thiocyanates are easily formed. The relative donor properties of solvent molecules and of competitive ligands are decisive for the formation of complex compounds.

[34] GUTMANN, V., and U. MAYER: to be published in Mh. Chem.
[35] BERNAL, I., and P. H. RIEGER: Inorg. Chem. **2**, 256 (1963).
[36] KIVELSON, D., and S. K. LEE: J. Chem. Phys. **41**, 1896 (1964).
[37] CARLIN, R. L., and F. A. WALKER: J. Amer. Chem. Soc. **87**, 2128 (1965).
[38] JONES, M. M., B. J. Yow, and N. R. MAY: Inorg. Chem. **1**, 166 (1962).

Index

Acceptor solvents, definition 4
— —, general 12
— —, proton-containing 59 ff.
— —, proton-free 80 ff.

Acetamide, as solvent 51 ff.
—, in liquid ammonia 37
—, in water 37
—, physical properties 52
—, reactions in 52 f.

Acetic acid, as solvent 36, 37, 53 ff.
— —, physical properties 53 f.
— —, reactions in 55 f.
— anhydride, as solvent 129 ff.
— —, physical properties 129
— —, reactions in 130 f.

Acetone, as solvent 144 ff.
—, physical properties 145
—, reactions in 146

Acetonitrile, as solvent 28, 131 ff.
—, physical properties 131
—, reactions in 132 ff.

Acetyl chloride, as solvent 27, 105, 107, 108

— —, physical properties 129
— —, reactions in 130 f.
— —, solvates of 107

Acidic function 5 ff.
— —, classical definition 5 f.
— —, ionotropic definition 8 f.
— —, Lewis concept 9
— —, protonic concept 6 f.
— —, solvent-system concept 7 f.

Alcohols, as solvents 38, 56 ff.
—, physical properties 57, 58
—, reactions in 57 f.

Aluminium(III) bromide, as solvent 98 ff.
— —, physical properties 99
— —, reactions in 101
— compounds, in various solvents 7, 26, 29, 50 ff., 56 f., 81, 84, 89, 96 f., 100 f., 106 ff., 112, 114, 116, 118 ff., 125, 127 ff., 132 f., 135 f., 146 ff., 151 ff.

Amides, as solvents 38
Amines, in various solvents 40, 50 ff., 54, 58 f., 66, 68, 81, 84, 109 f., 114

Ammonia (liquid), as solvent 34, 36, 37, 38 ff.
— —, physical properties 39 ff.
— —, reactions in 42 ff.
Ammonium compounds, in various solvents 39 ff., 50, 71, 73, 77, 84, 95
Ammonolysis reactions 42 ff.
Anionotropic solvent systems 8 f.
Antimony(III) chloride, as solvent 94
— —, physical properties 94
— —, reactions in 96 f.
Antimony-compounds, in various solvents 7, 14, 26, 29, 31, 48, 51 f., 56 f., 62, 67, 78, 81 f., 88 ff., 92, 94, 96, 98, 106 ff., 112, 114 f., 118 ff., 123 ff., 127 ff., 135 ff., 140, 146, 151 f., 158, 160 f., 168

Arsenic(III) bromide, as solvent 98 ff.
— —, physical properties 99
— —, reactions in 100
Arsenic(III) chloride, as solvent 94 f.
— —, physical properties 94 f.
— —, reactions in 96, 97
Arsenic compounds, in various solvents 48, 50 f., 56, 62, 66, 78, 83, 89 ff., 93, 95 ff., 100, 112, 128 f., 131, 135, 140, 145
Arsenic(III) fluoride, as solvent 92 ff.
— —, physical properties 93
— —, reactions in 93 f.

Autocomplex formation 24 f., 30 f., 115 f., 138, 143 ff., 150, 153 f., 157 f., 162 f., 165 ff.

Autofluoridolysis in fluorine-containing solvents 7, 9, 92

Autoprotolysis, of acetic and formic acids 36, 54
— — ammonia 35, 39, 40
— — ethyl- and methylalcohol 36, 37
— — fluorosulphuric acid 77
— — hydrazine 36
— — hydrogen chloride 36
— — hydrogen cyanide 36, 68
— — hydrogen fluoride 9, 35, 61
— — hydrogen sulphide 36, 49
— — nitric acid 36, 76
— — phosphoric acid 76
— — sulphuric acid 36, 70
— — water 6, 35

Azide bridging 167
Azide-ions, as competitive ligands 139, 143, 149, 150, 156, 166 f.
Azido-complexes, of transition metal ions 139, 143 f., 149, 150, 156, 166 f.

Barium Compounds, in various solvents 40, 62, 71, 79, 91, 95, 160
— Ion, half ware potentials in different solvents 33
Basic function 5 ff.
— —, classical definition 5 f.
— —, ionotropic definitions 8 f.
— —, Lewis concept 9
— —, protonic concept 6 f.
— —, solvent-system concept 7 f.
Benzoyl bromide, as solvent 28
— chloride, as solvent 27
— —, solvates of 107
— halides, as solvents 105, 107, 108 f., 110
— —, physical properties 105
— —, reactions in 110
Beryllium compounds, in various solvents 89, 95, 111, 148
Bismuth compounds, in various solvents 48, 50, 88 f., 107, 135, 140, 154, 158, 161 f.
Boron compounds, in various solvents 29, 40, 43, 50 f., 56 f., 61 f., 65 ff., 72 f., 76, 79 f., 83, 90 f., 106 ff., 112, 116 ff., 125, 132 f., 135, 141, 146 ff., 151, 158
Born-Haber cycle, of ion-solvation 32
Bromides, covalent, as solvents 98 ff.
Bromide-ion transfer reactions 99, 110
Bromide-ions, as competitive ligands 139, 143, 149 f., 156 f., 162 f.
Bromine(III) fluoride, as solvent 26 f., 87 ff.
— —, coordination compounds of 89
— —, physical properties 87
— —, reactions in 98 ff.
— —, self-ionization of 7
Bromo-complexes 83, 98 ff., 110, 137, 143, 145 f., 149 f., 156 f., 162 f.
Brönsted-Lowry theory 6, 36

Cadmium compounds, in various solvents 49, 51 f., 56, 79, 83, 100, 108, 110, 128, 134 f., 135, 141, 146 ff., 151, 158, 161
Calcium compounds, in various solvents 57, 88, 91, 95, 128, 160
— ion, half wave potentials in various solvents 33
Carbohydrates, in various solvents 40, 64

Carbonyl chloride, as solvent 104, 107, 108
— —, physical properties 105
— —, solvates of 107
— compounds, in various solvents 47, 74 f., 140
Carboxylic acids, as solvents 38
Cationotropic solvent systems 8 f.
Cesium compounds, in various solvents 52, 65, 84, 95, 100
— ion, half ware potentials in different solvents 33
Charge effect of metal chlorides in acetonitrile 135
Chlorides, covalent, as solvents 94 ff.
Chloride-ion transfer reactions 94, 96, 106, 113, 114, 117, 122
Chloride-ions, as competitive ligands 139 f., 143 ff., 163 ff.
Chlorine bridges 88, 105, 106, 133, 135
Chlorine(III) fluoride, as solvent 87 ff.
— —, physical properties 87
— —, reactions in 89
Chloro-complexes 26, 64 ff., 82, 94 ff., 107 ff., 111 ff., 128 ff., 133 ff., 139 f., 143 ff., 163 ff.
Chlorosulphuric acid, as solvent 77 ff.
— —, physical properties 77
— —, reactions in 79
Chromium compounds, in various solvents 29, 40, 44, 47, 74, 76, 89, 91, 100, 110, 136 f., 139 f., 143, 149 f., 153, 156 ff., 165 ff.
Class (a) ions 10
Class (a) metal complexes, formation 28, 30
— — — —, stability 161, 162, 163 f.
Class (b) ions 10
Class (b) metal complexes, formation with SO_2 81
— — — —, stability 161 f., 163
Cobalt compounds, in various solvents 29, 31, 40, 43, 45 ff., 48 f., 51, 64, 79, 110, 128, 137 ff., 141 ff., 145, 148 ff., 153 ff., 159 ff., 164, 166 f.
Colour indicators 50, 55, 58, 68, 109, 123 f., 130
Coordination chemistry in donor solvents of transition metal ions 161 ff.
— — — proton-containing donor solvents 35 ff.
— equilibria in solution 24 ff.
Copper compounds, in various solvents 51 f., 75, 79, 100, 107, 138 ff., 149 f., 152 f., 156 f., 160, 165 f.
Covalent bromides, as solvents 98 ff.
— chlorides, as solvents 94 ff.
— fluorides, as (proton-free) acceptor solvents 86 ff.

Covalent
— oxides, as (proton-free) acceptor solvents
  80 ff.
Cyanide-ions, as competitive ligands 143,
  156 f., 167 f.
Cyanides 68 f., 167
Cyano-complexes, of transition metal ions
  46, 143, 156 f., 167 f.

1,2-Dichloroethane, as solvent 127
—, physical properties 127
—, reactions in 127
Dielectric constant, as important factor for
  an ionizing solvent 4 f.
Diethylether, as solvent 147 f.
—, physical properties 147
—, reactions in 147 f.
Difluorophosphoric acid, as solvent 77 f.
— —, physical properties 77
— —, reactions in 79
N,N-Dimethylacetamide, as solvent 154
—, physical properties 154
—, reactions in 154
Dimethylformamide, as solvent 34, 152 ff.
—, physical properties 152
—, reactions in 153 f.
Dimethyl sulphoxide, as solvent 30, 31, 34,
  155 f.
—, physical properties 155
—, reactions in 156 ff.
Dinitrogen tetroxide, as solvent 84 ff.
— —, complex compounds of 84 f.
— —, physical properties 84
— —, reactions in 85 f.
Dissociation, of solvate complexes 28 ff.
Disulphuric acid, as solvent 77 ff.
— —, physical properties 77
Donor-acceptor-interactions, enthalpies of
  13 ff.
— — —, prediction of 20 f.
Donor number 19 f., 103
— —, application 25 f., 29 f., 31
— —, attempted assignment to anions 168
— —, definition 14
— —, relation to solvation 32
— properties of solvents 31 ff.
— solvents 12, 128
— —, coordination chemistry of transition
  metal ions in 161 ff.
— —, definition 4
— —, general 128
— —, oxyhalides as 103 ff.
— — (proton-containing), coordination chemistry in 35 ff.
— strength 12 ff.
— —, definition 14

Enthalpies of adduct formation 15 ff.
Esters 40, 146
Ethylacetate, as solvent 146
—, physical properties 146
—, reactions in 146

Fluoride bonding, in proton-free solvents 86,
  87, 92
Fluoride-ion transfer reactions 9, 62, 86, 89,
  92
Fluorides, covalent, as solvents 86 ff.
Fluoro-complexes 61 f., 78 f., 83, 89 ff., 155 f.
Fluorosulphuric acid, as solvent 77 ff.
— —, physical properties 77
— —, reactions in 78 f.
Formamide, as solvent 51 ff.
—, physical properties 52
—, reactions in 52 f.
Formic acid, as solvent 36, 53 ff.
— —, physical properties 53 f.
— —, reactions in 52 f.

Gallium compounds, in various solvents 100,
  106 f., 109 f., 112, 123, 127, 132 f., 135,
  147
Germanium compounds, in various solvents
  47, 62, 75, 89 ff., 132, 135
Gold compounds, in various solvents 79,
  88 ff., 99 f., 107, 110, 112

Hafnium compounds, in various solvents 112
Half-wave potentials of group Ib- and IIb-
  ions in different solvents 32 f.
Hard acids and bases 4, 10
Hexamethylphosphoramide, as solvent 159f.
—, physical properties 159
—, reactions in 160
Hydrazine, as solvent 36, 37, 38, 48 ff.
—, physical properties 48 f.
—, reactions in 49
Hydrocarbons, in various solvents 40, 50
Hydrogen bonding, in proton-containing
  solvents 35, 40, 48, 49, 54, 57, 60, 69,
  71, 141, 152
— —, in solvents 21 ff., 35
— bromide, as solvent 64 ff.
— chloride, as solvent 64 ff.
— cyanide, as solvent 36, 38, 67 ff.
— —, physical properties 67
— fluoride, as solvent 26 f., 36, 37, 60 ff.
— —, physical properties 60
— —, reactions in 61 ff.

Hydrogen
— halides (except HF), as solvents 38, 64 ff.
— — —, physical properties 64
— iodide, as solvent 66
— sulphide, as solvent 36, 49 ff.
— —, physical properties 50
— —, reactions in 50 f.

Indium compounds, in various solvents 49,
    107, 110, 112, 125, 132, 134 f.
Iodide-ion transfer reactions 101
Iodide-ions as competitive ligands 139, 143,
    150, 156, 159, 161 f.
Iodine, as solvent 101 f.
—, in various solvents 18, 50, 74, 97, 158
—, physical properties 101
—, reactions in 102
— monobromide, as solvent 28, 98 ff.
— —, physical properties 99
— —, reactions in 100
— monochloride, as solvent 65, 94 f.
— —, physical properties 94
— —, reactions in 95 f., 97
— pentafluoride, as solvent 91 f.
— —, physical properties 91
— trichloride 88
Iodo-complexes 83, 102, 137, 139 ff., 143,
    145, 150, 156, 159, 161 f.
Ionization 24
— of solvate complexes 24, 28 ff.
Ionotropic definitions of acids and bases 8 f.
Iridium compounds, in various solvents 47
Iron compounds, in various solvents 13, 27,
    29 ff., 47, 50 ff., 56, 69, 74 f., 79, 83, 86,
    96 f., 107 ff., 111 f., 114 ff., 117 ff., 125,
    128, 130, 139, 143 f., 146, 148, 150 ff.,
    154 ff., 159 f., 165, 167

Lanthanum compounds, in various solvents 48
Lead compounds, in various solvents 40, 42,
    46, 52, 61, 74, 100, 107, 135, 154, 158, 161
Ligand effect, for metal chlorides in acetoni-
    trile 135
— exchange, general 24
Lithium compounds, in various solvents 39,
    49, 52, 71, 77, 85, 95, 131
— ion, half wave potential in different sol-
    vents 33

Magnesium compounds, in various solvents
    85, 89, 95, 146, 160
— ion, half wave potential in different sol-
    vents 33

Manganese compounds, in various solvents
    29, 43, 47 f., 51, 57, 74, 79, 85, 89 ff.,
    107, 125, 139 f., 142 ff., 145, 150 f., 156,
    160, 162 ff.
Mercuric bromide, as solvent 28, 98
— —, physical properties 99
— —, reactions in 100
Mercury compounds, in various solvents 7,
    29, 48, 51 f., 69, 75, 83, 100 ff., 107 f., 110,
    120 ff., 131, 133 ff., 140, 154, 159, 162
Molybdenum compounds, in various solvents
    43 f., 83, 112, 118, 130, 137, 146, 153
Molecular liquids, as solvents 2
Molten metals, as solvents 2
— salts as solvents 2

Neptunium, complex halides in acetonitrile
    137 f.
Nickel compounds, in various solvents 29,
    40 f., 43, 46, 49, 52, 79, 110, 128, 138 ff.,
    142 ff., 145, 148 ff., 153, 156 f., 159 ff.,
    164, 166 f.
Niobium compounds, in various solvents 29,
    44, 56, 62, 83, 88 ff., 94, 107, 110, 112, 118,
    121, 129, 135, 137, 147, 158
Niobium(V) fluoride, as solvent 86
Nitric acid, as solvent 36, 37, 38, 75 f.
— —, physical properties 75
— —, reactions in 76
Nitrobenzene, as solvent 127 ff.
—, physical properties 127
Nitrogen dioxide, see Dinitrogen Tetroxide
Nitromethane, as solvent 127 ff.
—, physical properties 127
—, reactions in 128
Nitrosyl bromide, as solvent 28
— chloride, as solvent 27, 105 ff.
— —, physical properties 105
— —, solvates of 107
— —, structure 106
Non-aqueous solvents, classification 1 ff.
— —, general 1 ff.
— —, principles of coordination chemistry in
    12 ff.
— —, techniques 11
Non-ionizing solvents 2
Non-solvated complex anions 25 ff.

Onium complexes 24, 29, 121
Osmium, complex fluorides of 88
Oxides, covalent, as (proton-free) acceptor
    solvents 80 ff.
Oxidation states in liquid ammonia 46 f.

Oxyhalides as solvents, with low donor numbers 104 ff.
— — —, with medium donor numbers 111 ff.
—, donor numbers of 103

Palladium compounds, in various solvents 46, 48, 107, 155
Perchloric acid, as solute 5, 37, 55, 63, 71, 79, 131
Phenol, in various solvents 18, 40, 56
Phenylphosphonic dichloride, as solvent 27, 111
— —, physical properties 111
— —, reactions in 113 ff.
Phosphoric acid, as solvent 38, 75 f.
Phosphorus compounds, in various solvents 29, 42, 50 f., 62, 65 f., 72, 74, 76, 79, 82, 89 ff., 93 ff., 97 ff., 108 ff., 118, 121 ff., 125, 129 f., 135 f., 141
Phosphorus oxychloride, as solvent 27, 31, 111
— —, physical properties 111
— —, reactions in 113 ff.
Platinum compounds, in various solvents 47 f., 79, 89, 91, 107
Plutonium(IV), complex fluorides of, in acetonitrile 137 f.
Potassium compounds, in various solvents 39, 49, 52, 54, 57, 61 f., 68, 71, 74, 76 f., 83 ff., 88 ff., 95, 97 f., 100 ff., 110, 159
— ion, half wave potential in different solvents 33
Principles of coordination chemistry in non-aqueous solvents 12 ff.
Propanediol-1,2-carbonate, as solvent 28, 31, 142 ff.
— —, physical properties 142
— —, reactions in 143 f.
Protactinium(IV), complex fluorides of, in acetonitrile 137 f.
Proteins, in HF 64
Proton-containing acceptor solvents 59 ff.
— — donor solvents, coordination chemistry in 35 ff.
— — — —, general properties 35 ff.
— — solvents 3, 4, 35 ff., 59 ff.
Proton-free acceptor solvents 80 ff.
— — — —, covalent bromides 98 ff.
— — — —, covalent chlorides 94 ff.
— — — —, covalent fluorides 86 ff.
— — — —, covalent oxides 80 ff.
— — — —, iodine 101 f.
— — solvents, general 3, 4
Proton-transfer reactions (protolysis) 6 f., 35 f., 39, 49, 50, 52, 53, 57, 60 f., 71, 78, 79

Rhenium compounds, in various solvents 90, 110
Rubidium compounds, in various solvents 77, 88, 91, 95
— ion, half wave potential in different solvents 33
Ruthenium(V), complex fluorides of, in BrF₃ 89 ff.

Selenium compounds, in various solvents 29, 46, 51, 72, 74, 79, 121
— oxychloride, as solvent 27, 111
— —, physical properties 111
Self-ionization equilibria 35
— — aluminium bromide 99
— — (liquid) ammonia 7, 39, 40
— — antimony chloride 94
— — arsenic bromide 99
— — arsenic chloride 94
— — arsenic fluoride 92
— — bromine trifluoride 7, 9, 27
— — carbonyl chloride 7
— — hydrogen fluoride 9, 27, 60 f.
— — hydrogen halides (others than HF) 65
— — iodine(V) chloride 92
— — iodine(V) fluoride 92
— — mercury(II) bromide 7, 99
— — nitrosyl chloride 27, 104
— — phosphorus oxychloride 113, 115
— — selenium oxychloride 113
— — solvents (cf. also autoprotolysis etc.) 6 ff., 27
Silicon compounds, in various solvents 43, 73, 89, 91, 93, 107, 135
Silver compounds, in various solvents 40 f., 48, 50, 61 ff., 69, 71, 75, 84 f., 85 ff., 100, 108, 140, 148, 153
Size effect of metal chlorides in acetonitrile 135
Sodium, in solution 45 f., 51, 168
— compounds, in various solvents 39, 42, 45, 47, 49, 52, 54, 57, 61, 71, 75, 77, 85, 88, 97, 100, 141, 148
— ion, half wave potential of, in different solvents 33
Soft acids and bases 4, 10
Solubilities, in solvents 23 f., 39, 48, 50, 61, 80, 88, 148
Solvate complex, ionization and dissociation 28 ff.
Solvation 13, 31 ff.
— number 34

Solvents, classification of 1 ff.
—, coordinating properties of 12 ff.
—, donor number of 19 f.
—, donors 126 ff.
—, hydrogen bonding in 21 ff.
—, non-aqueous, classification 2
—, — —, definition 1 f.
—, — —, principles of coordination chemistry in 12 ff.
—, non-coordinating 3
—, physical properties 4 f.
—, proton-containing, acceptors 59 ff.
— — —, donors 35 f.
Solvent-system concept of acids and bases 7 f.
Stannic chloride, as solvent 3
Steric factors, contributing to coordination chemistry in solvents 23, 142, 151, 159 f.
Strontium compounds, in various solvents 61, 71, 95
— ion, half wave potential in different solvents 33
Sulpholane, as solvent 28, 141 f.
—, physical properties 141
—, reactions in 142
Sulphur compounds, in various solvents 43, 46, 73, 75, 88 f., 141
— dioxide, as solvent 80 ff.
— —, physical properties 80
— —, reactions in 83 f.
Sulphuric acid, as solvent 36, 37, 38, 69 ff.
— —, physical properties 69, 70 f.
— —, reactions in 71 ff.
Sulphuryl chloride, as solvent 27, 107, 108
— —, physical properties 105

Tantalum compounds, in various solvents 29, 44, 56, 62, 79, 88 ff., 108, 112, 118, 121, 135, 137
Tantalum(V) fluoride, as solvent 86
Techniques with non-aqueous solvents 11
Tellurium compounds, in various solvents 46 f., 62, 79, 96, 108, 112, 130, 134 f., 146
Thallium compounds, in various solvents 61, 100, 107, 112, 135, 137, 145, 148
Thiocyanate-ions as competitive ligands 139, 143, 167
Thiocyanato-complexes 139, 143, 167
Thionyl chloride, as solvent 27, 107, 108
— —, physical properties 105

Thionyl chloride,
— —, reactions in 84, 109 f.
— —, solvates of 107
Thorium compounds, in various solvents 56, 107
Tin compounds, in various solvents 17 f., 27, 29, 50 ff., 56 f., 67, 74, 79, 81, 88 ff., 96, 107 ff., 112, 118 ff., 125, 127, 129 f., 132 f., 135 ff., 140, 151 f., 154 f., 158, 160 ff.
Titanium compounds, in various solvents 29, 43 ff., 50 f., 56 f., 79, 81, 89, 96 ff., 100, 106 ff., 112, 117 ff., 125 f., 132 f., 135 ff., 139 f., 143 f., 146, 150 f., 155, 165 f.
Transition metal ions, coordination chemistry in donor solvents 161 ff.
Transport numbers of cations in H_2SO_4 71
Tributyl phosphate, as solvent 28, 31, 148 ff.
—, physical properties 149
—, reactions in 149 ff.
Triphenylcarbonium-ion 66, 81, 96, 117
Triphenylchloromethane, as chloride ion donor 12, 25, 28, 56, 59, 66, 81, 96, 117, 122, 133
Tungsten compounds, in various solvents 83, 93, 110, 112, 118, 121, 130, 137 f., 140

Uranium compounds, in various solvents 56, 62, 76, 83, 107, 138

Vanadium compounds, in various solvents 29, 43 f., 74, 89 ff., 96 f., 107, 136 f., 139 f., 142 f., 149 f., 156, 162 f., 165 f.
Vanadium(V) fluoride, as solvent 86
Vanadyl compounds, in various solvents 22, 29 ff., 74, 128, 139 f., 143 f., 149 f., 155 ff., 166, 168

Xenon, fluorides of, in HF 62 f.

Zinc compounds, in various solvents 29, 40, 48 f., 52, 56, 79, 86, 89, 107 ff., 110, 118 ff., 133 f., 140, 151, 160
Zirconium compounds, in various solvents 29, 44, 56, 81, 97, 107 f., 112, 121, 128, 132, 135, 146